Analytical instrumentation for the water industry

Analytical instrumentation for the water industry

T. R. CROMPTON, BSc, MSc

Series Editor: B. E. NOLTINGK

Butterworth-Heinemann Ltd
Linacre House, Jordan Hill, Oxford OX2 8DP

 PART OF REED INTERNATIONAL BOOKS

OXFORD   LONDON   BOSTON
MUNICH   NEW DELHI   SINGAPORE   SYDNEY
TOKYO   TORONTO   WELLINGTON

First published 1991

British Library Cataloguing in Publication Data
Crompton, T. R.
   Analytical instrumentation for the water industry.
   I. Title
   628.1028

ISBN 0 7506 1139 1

Library of Congress Cataloguing in Publication Data
Crompton, T. R. (Thomas Roy)
   Analytical instrumentation for the water industry/T.R. Crompton
      p.   cm.
   Includes bibliographical references and index.
   ISBN 0 7506 1139 1
      1. Water—Analysis.   2. Water quality—Measurement—Equipment and
   supplies.   I. Title.
   TD380.C76   1991
   628.1'61—dc20   91–15642
                                                          CIP

Composition by Genesis Typesetting, Laser Quay, Rochester, Kent
Printed and bound in Great Britain by Billings Ltd., Worcester

Contents

Preface

Great strides have been made in recent years in the analytical instrumentation made available in water laboratories, both chemical and biological and in chemical plant associated with the water industry. It is the purpose of this book to describe the types of instrumentation now available to the water chemist and engineer, and to discuss their capabilities and applications. It is hoped that the book will be of interest to those responsible for planning and running laboratories and works. All aspects of modern instrumentation are covered, including those used in general quality control, effluent analysis monitoring routines, accidental spillages and radioactivity and biological monitoring. Aspects such as automated analysis, robotics, computerized control of instruments, laboratory management systems, quality control routines and safety in the laboratory and the plant are included in the discussion. The book is intended therefore, for all staff, whether direct or peripheral, who are concerned with instrumentation in the water laboratory, including laboratory designers, engineers, chemical engineers and those concerned with the implementation of directives or legal requirements.

It is also hoped that the book will be of interest to students not only those who are interested in water technology, but those who have an interest in laboratories or process control in general.

Acknowledgements

Many companies actively provided assistance to the author during the preparation of the manuscript by providing information regarding their products. Many thanks are due to these companies who are too numerous to mention here but recognition is given where relevant in the text, and the companies, their locations and their products are listed in the Appendix.

Thanks are due to Mr Chris Lee, Safety Officer of the North West Water Authority, for his excellent and timely review in Chapter 11, on on-site safety instrumentation.

The author would like to record his appreciation to Mrs Pam Green, who completed the typing with speed and efficiency, and also to his wife, Elisabeth, who provided much assistance.

T. R. Crompton

1 Introduction

During the past twenty years great strides have been made in analytical chemistry technology. Developments such as graphite furnace atomic absorption spectrometry, inductively coupled plasma atomic emission spectrometry and the combination of the inductively coupled plasma technique with mass spectrometry have ensured great steps forward in the capabilities of all analytical laboratories including those in the water industry.

During the same time span, for various reasons the demands made on analytical laboratories in the water industry have increased tremendously in terms of the number of samples requiring analysis, in the need for improved sensitivity and in the increase in the number of substances that have to be analysed for. The increase in the range of interests and responsibilities of such laboratories has been brought about by the necessity to implement EEC and other directives, the increasing concern for the environment and new responsibilities such as coastal waters.

Fortunately the improvement in analytical abilities has coincided with the increase in demands, so that the modern water laboratory has been able to keep pace with the demands made of it.

Water laboratories must, of course, analyse water samples and effluents, i.e. natural waters, rivers, ponds, ground waters, boreholes, potable water, rainwater, sewage and trade effluents and wastewaters. They must also analyse non-aqueous samples, such as river and ocean bed sediments, fish, crustacea and plant material.

In all cases sample numbers are large, and in many cases sensitivity requirements are high. Thus, whereas a few years ago analysis would be required at the $mg\,l^{-1}$ level, nowadays, $\mu g\,l^{-1}$ analysis is commonplace and, frequently, analysis is required at the sub $\mu g\,l^{-1}$ level.

An increasing area of responsibility has been the monitoring of rivers, water supply inlets, potable waters, rainwater and seawater for traces of organic and inorganic substances of interest from the point of view of human health. In particular, the water laboratory has to be ready at very short notice to identify and determine constituents resulting from accidental discharges into rivers and seawater. Here, the most refined techniques are required, such as a combination of gas chromatography and mass spectrometry for organics and a combination of inductively coupled plasma atomic emission spectrometry and mass spectrometry for metals.

The water laboratory must also monitor substances that are released more slowly into the aqueous environment, such as insecticides and nitrates, percolating into rivers from farm land.

The incidence of acid rain is of increasing environmental concern: traces of anions build up in rainwater with dramatic effects on inland water ecology. Here, ion chromatography with its ability to carry out multi-component analysis of very low levels of anions is proving to be of great interest.

The incidence of accidents involving nuclear power stations, such as Chernobyl, has meant that water laboratories have had to take on board certain responsibilities for monitoring radioactivity in water and sewage effluent samples. Radioactivity laboratories are now being set up for this purpose for the measurement of both α, β and γ activity and also of individual radioisotopes.

The range of activities of biological laboratories in the water industry is also increasing. Not only that, but larger numbers of samples require analysis, hence fully automated immunoassays are necessary in many cases.

In order to lighten the load on the analytical laboratory and at the same time to enable them to make better decisions on the spot, river inspections are increasingly using on-site instrumentation to measure parameters such as pH, electrical conductivity, dissolved oxygen, anions and cations. Appropriate use of these techniques on site enables the inspector to take more meaningful samples for more detailed laboratory examination.

Vast improvements have also been made in recent years in on-line analytical monitoring equipment for use in water treatment plants, sewage works, etc. Continuous monitoring with telemetry, if needed, is now possible for disinfection gases (chlorine, ozone), treatment chemicals (fluoride), operating parameters (EC, pH, dissolved oxygen, turbidity) and cation and anion levels.

Another rapidly developing area is that of safety instrumentation for use in on-site gas testing for flammable and toxic substances in sewers and sewage digesters. Gases that can now be determined include oxygen, carbon dioxide, carbon monoxide, methane and/or higher paraffins, helium, hydrogen, hydrogen sulphide, sulphur dioxide and chlorine.

In general, the most sensitive methods of analysis are required for potable water, river, estuary and sea water and for these the more recent highly sensitive methods are required. These include for metals graphite atomic absorption spectrometry, Zeeman atomic absorption spectrometry and inductively coupled plasma atomic absorption spectrometry. For organics, the method of choice is gas chromatography or high performance liquid chromatography coupled with mass spectrometry, which combines high sensitivity with the ability to identify unknown substances.

When low levels of unknown metals are being sought techniques such as neutron activation analysis, X-ray fluorescence spectrometry and inorganic mass spectrometry come into these areas.

Generally speaking, sewage effluents and industrial effluents contain higher concentrations of metals and organics and, in many instances, less demanding methods of analysis may suffice.

The determination of volatile substances in water samples and effluents is a special case and to determine these headspace and purge and trap techniques will be required.

Increasingly, inspectors concerned with sewer pollution, industrial effluents and accidental spillages are carrying test kits so that measurements can be performed rapidly on-site. This permits more intelligent selection of locations at which to take samples for more detailed laboratory examination later. It also enables them to issue local warnings more quickly, such as those to local farmers to keep their cattle away from river banks if a spillage has occurred.

Analytical instrumentation is being installed in water treatment plants and upstream of raw water intake points on river banks, again to obtain the earliest possible indication of pollution and to generally assist in process control.

Many regulatory bodies now exist who are concerned with the quality of potable water and river and sea water and also the composition of sewage sludge disposed of to land. In general, these bodies stipulate maximum levels for metals and organics and do not insist on particular analytical techniques being used. One of the purposes of this book is to assist the authorities in the water industry to choose the technique most appropriate to their needs.

2 Determination of metals

2.1 Spectrometric techniques – general discussion

2.1.1 Atomic absorption spectrometry (AAS)

Since shortly after its inception in 1955 atomic absorption spectrometry has been the standard tool employed by analysts for the determination of trace levels of metals in water samples. In this technique a fine spray of the analyte is passed into a suitable flame, frequently oxygen acetylene, or nitrous oxide acetylene, which converts the elements to an atomic vapour. Through this vapour is passed radiation at the right wavelength to excite the ground state atoms to the first excited electronic level. The amount of radiation absorbed can then be measured and directly related to the atom concentration: a hollow cathode lamp is used to emit light with the characteristic narrow line spectrum of the analyte element. The detection system consists of a monochromator (to reject other lines produced by the lamp and background flame radiation) and a photomultiplier. Another key feature of the technique involves modulation of the source radiation so that it can be detected against the strong flame and sample emission radiation.

The technique can determine a particular element with little interference from other elements. It does however have two major limitations. One of these is that the technique does not have the highest sensitivity, and the other is that only one element at a time can be determined. This has reduced the extent to which it is currently used.

2.1.2 Graphite furnace AAS

The graphite furnace atomic absorption technique, first developed in 1961 by L'vov, is an attempt to improve the detection limits achievable in water analysis. In this technique, instead of being sprayed as a fine mist into the flame, a measured portion of the sample is injected into an electrically heated graphite boat or tube, allowing a larger volume of sample to be handled. Furthermore, by placing the sample on a small platform inside the furnace tube, atomization is delayed until the surrounding gas within the tube has heated sufficiently to minimize vapour phase interferences, which would otherwise occur in a cooler gas atmosphere.

The sample is heated to a temperature slightly above 100°C to remove free water, then to a temperature of several hundred degrees centigrade to remove water of fusion and other volatiles. Finally, the sample is heated to a temperature near to 1000°C to atomize it and the signals thereby produced are measured by the instrument.

The problem of background absorption in this technique is solved by using a broad-band source, usually a deuterium arc or a hollow cathode lamp, to measure background independently and subsequently to subtract it from the combined atomic and background signal produced by the analyte hollow cathode lamp. By interspersing the modulation of the hollow cathode lamp and 'background corrector' sources, the measurements are performed apparently simultaneously. Specific advances have been achieved such as optical control of furnace temperatures to improve heating rates and hence measurement sensitivities. Better furnace tube geometries and gas-flow systems have helped to reduce interference effects. Methods have been devised of mass producing pyrolytic graphite coatings on electrographite tubes and this has reduced the porosity of atom containment and stable carbide formation.

Graphite furnace techniques are about one order of magnitude more sensitive than direct injection techniques. Thus lead can be determined down to $50\,\mu g\,l^{-1}$ by direct AAS and down to $5\,\mu g\,l^{-1}$ by GFAAS.

In recent years, even greater demands for increases in sensitivity have been placed on water chemists. This has raised the popularity of the furnace technique. In turn, it has placed even greater demands on background correction systems. Great improvements have been made in continuum sources methods, and additionally two alternative techniques have reached commercial fruition. The first is based on measuring the background during a high current pulse of the source hollow cathode lamp, thereby broadening the lamp-emitted line profile. The second is by using Zeeman or magnetically induced splitting of either the hollow cathode lamp emission profile or the sample adsorption profile. Both techniques thus allow the measurement of background levels without using a second source.

2.1.3 Zeeman AAS techniques

The Zeeman technique, though difficult to establish, has an intrinsic sensitivity perhaps five times greater than that of the graphite furnace technique, say $1\,\mu g\,l^{-1}$ detection limit for lead.

Non-specific background attenuation has always been the most common type of interference in graphite furnace atomic absorption spectrometry. It is caused by higher concentrations of molecular species, small droplets, salt particles or smoke which may absorb or scatter the light emitted from the primary light source.

Most atomic absorption spectrometers for use with the graphite furnace technique are therefore equipped with continuum source background correctors which automatically compensate for broad band absorption interferences. These are sufficient for many routine applications. Background correction with a continuum light source is however limited to background absorption which is 'uniform' within the spectral bandwidth. In addition, background signals can be corrected only up to 0.7 absorbance units.

The Zeeman effect is exhibited when the intensity of an atomic spectral line, emission or absorption, is reduced when the atoms responsible are subjected to a magnetic field, nearby lines arising instead (Figure 2.1). This makes a powerful tool for correction of background attenuation caused by molecules or particles, which does not normally show such an effect. The technique is to subtract from a 'field-off' measurement the average of 'field-on' measurements made just beforehand and just afterwards. See Figure 2.2.

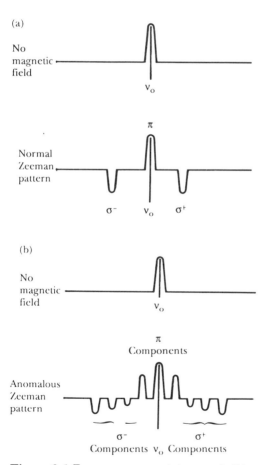

Figure 2.1 *Zeeman patterns: (a) normal; (b) anomalous*

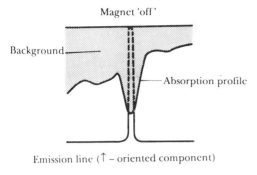

Magnet 'off'

Background

Absorption profile

Emission line (↑ – oriented component)

(a)

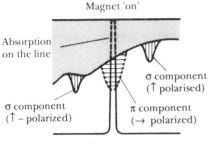

Magnet 'on'

Absorption on the line

σ component
(↑ polarised)

σ component
(↑ – polarized)

π component
(→ polarized)

Emission line (↑ oriented component)

(b)

Figure 2.2 *Zeeman patterns: (a) analyte signal plus background; (b) background only*

The simultaneous, highly resolved graphic display of the analyte and the background signals on the video screen provides a means of reliable monitoring of the determination and simplifies methods development.

Besides background attenuation chemical interferences can be a major limitation in the practical use of graphite furnace atomic absorption spectrometry. Chemical interferences in furnace atomic absorption spectrometry often lead to strong signal suppression. This cannot be removed by any type of background correction. The way to reduce or completely eliminate these matrix interferences is to atomize samples into the environment that has almost reached thermal equilibrium. Under such conditions the formation of free atoms is optimum and recombination of atoms to molecules or loss of atoms is effectively avoided.

The stabilized temperature platform furnace eliminates chemical interferences to such an extent that in most cases personnel and cost-intensive sample preparation steps, such as solvent extractions, as well as the time-consuming method of additions are no longer required.

The advantages of Zeeman background correction are:

1 correction over the complete wavelength range
2 correction for structural background
3 correction for spectral interferences
4 correction for high background absorbances
5 single element light source with no possibility of misalignment

2.1.4 Preconcentration AAS techniques

Detection limits can be improved still further in the case of all three techniques mentioned above by use of a preconcentration technique. One such technique which has found great favour involves converting the metals to an organic chelate by reaction of a larger volume of sample with a relatively small volume of an organic solvent solution, commonly of diethyldithiocarbamates or ammonium pyrrolidone dithiocarbamates. The chelate dissolves in the organic phase and is then back-extracted into a small volume of aqueous acid for analysis by either of the techniques mentioned above. If 500 ml to 1 litre of sample is originally taken and 20 ml of acid extract finally produced then concentration factors of 25–50 are thereby achieved with consequent lowering of detection limits achievable. Needless to say, this additional step in the analysis considerably increases analysis time and necessitates extremely careful control of experimental conditions.

Microscale solvent extractions involving the extraction of 2.5 ml sample with 0.5 ml of an organic solvent solution of a chelate give detection limits for lead and cadmium by the Zeeman GFAAS method respectively, of 0.6 and 0.02 μg l^{-1}. Thus, for cadmium, analyses in the low ng l^{-1} range is becoming possible.

2.1.5 Microprocessors

In recent years the dominating influence on the design and performance of atomic absorption spectrometers is that of the microprocessor. Even the cheapest instruments are expected to provide autosampling systems for both flame and furnace use and therefore a means of recording the data produced. The most important feature characterizing the higher-cost instruments is that of sequential multi-element analysis without operator intervention. Instruments are now available which require only the insertion of the appropriate (hels) cassette and a predefined 'task' of various element determinations, may be performed under fully optimal conditions in either flame or furnace mode. It is possible for instruments to mimic the procedures of trained analysts by varying conditions such as flame composition, monochromator bandpass and wavelength automatically to ensure the best possible analysis parameters.

2.2 Flame and graphite furnace atomic absorption spectrometry

2.2.1 Instrumentation

Increasingly, due to their superior intrinsic sensitivity, atomic absorption spectrometers available are capable of implementing the graphite furnace techniques. Some available instrumentation in flame and graphite furnace atomic absorption spectrometry is listed in Table 2.1.

In Figure 2.3(a) and (b) are shown the optics of a single-beam flame spectrometer (Perkin-Elmer 2280) and a double-beam instrument (Perkin-Elmer Model 2380).

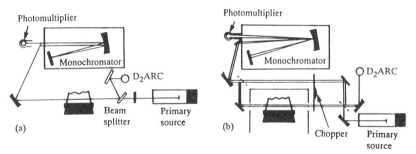

Figure 2.3 *(a) Optics Perkin-Elmer Model 2280 single beam atomic absorption spectrometer; (b) Optics Perkin-Elmer 2380 double beam atomic absorption spectrometer*

The model 2280 has a very simple optical system. Energy from the primary source is focused through a lens into the sample compartment and then to the monochromator. As only one lens is used, compensation for the wavelength influence on energy throughput is easily done by positioning the hollow cathode lamp along the horizontal axis.

In the double-beam Model 2380 light from the primary source is divided into two beams – a sample beam and a reference beam. The sample beam travels through the sample compartment, while the reference beam travels around it. The beams are recombined before entering the monochromator. The double-beam system compensates for any changes that may occur in lamp intensity during an analysis. The signal produced is actually a ratio between the two beams. Therefore, any fluctuations in the light output will affect both beams equally and will be compensated for automatically. This results in a more stable baseline and ultimately in better detection limits.

Table 2.1. *Available flame and graphite furnace atomic absorption spectrometers*

Type of instrument	Supplier	Model no. and type	Microprocessor	Hydride and mercury attachment	Autosampler	Wavelength range
Flame (direct injection)	Thermo-electron	1L 157 single channel 1L 357 single beam	–	Yes	Yes	
		1L 457 single channel double beam	with graphics			
		Video 11 single channel single beam	with graphics			
		Video 12 single channel double beam	with graphics			
		Video 22 two double channels	computer interference			
Graphite furnace	Thermo-electron	1L 655 CTF	–	–	Yes	
Direct injection	Perkin-Elmer	2280 single beam	Yes	–	–	190–870
		2380 double beam	Yes (with automatic background correction)	–	–	190–870
Graphite furnace	Perkin–Elmer	100 single beam	Yes	–	–	190–870
		2100 single path double beam	Yes	Yes	–	190–870

Type	Company	Instrument				Range
Graphite furnace	Varian Associates	SpectrA A30/40 multi-element analysis. Method storage	Yes	Yes	Yes	190–900
		SpectrA A10 (low cost, single beam)	Yes (built in VDU)	Yes	Yes	190–900
		SpectrA A20 (medium cost, double beam)	Yes (built in VDU)	Yes	Yes	190–900
Flame graphite furnace		SpectrA A 300/400 multi-element analysis, centralized instrument control	Yes (with colour graphics and 90 elements on disk)	Yes	Yes	190–900
		STA 9S and GTA 96 graphite tube atomizer units – compatible with all SpectrA A instruments	Furnace and programmable sample dispenser operated from SpectrA A keyboard. Rapid interchange between flame and furnace operation			
Graphite furnace	GBC Scientific Pty Ltd.	903 single beam	Yes	Yes	Yes	176–900
		902 double beam (both with impact head option)	Yes	Yes	Yes	170–900
Flame (direct injection) graphite furnace	Shimadzu	AA670 double beam	Yes	Yes	Yes	190–900
		AA670 G double beam	Yes	Yes	Yes	190–900

The light beam is transmitted through the system by front-surfaced, quartz-coated mirrors. The advantages of using reflecting optics are that the efficiency in energy throughout is unaffected by the wavelength being considered and no additional focusing optics are required. All mirrors used in the system are also specially coated with silicon oxide, which protects the surface if the instrument is operated in a corrosive laboratory atmosphere.

The monochromator uses a very finely ruled grating, the efficiency of which is dependent upon its area, its dispersion and its blaze angle. A dual-blazed grating is used in the Perkin-Elmer model 2380 with two blaze angles, one in the ultraviolet at 236 nm and the other in the visible range at 597 nm. This distributes energy throughput more evenly throughout the wavelength range (190–870 nm). The model 2280 uses a blazed grating with one blaze angle at 255 nm.

The burner assembly of the Perkin-Elmer models 2280 and 2380 instruments incorporates an impact bead which can lead to an appreciable improvement in detection limits achieved to those attainable with a conventional flow spoiler design.

The impact bead cannot be used for all applications. It is not recommended for the nitrous oxide–acetylene flame; the sensitivity improvement is not noticed here. Furthermore, due to minor flame instability, poorer precision and somewhat poorer detection limits will result. The bead also should not be used when analysing solutions containing high concentrations of dissolved salts as these solutions may clog the burner system.

The flow spoiler is selected when determining elements requiring the nitrous oxide–acetylene flame for best precision or when analysing solutions containing high concentrations of dissolved solids. In addition the flow spoiler is used when working with solutions which may corrode the impact bead and cause contamination problems.

The Perkin-Elmer burner system has the ability to use an auxiliary oxidant flow. This capability contributes to improved flame stability and precision and it provides a convenient and simple means of obtaining proper flame conditions when using combustible organic solvents.

The Perkin-Elmer models 2280 and 2380 spectrometers are available with several types of burner control systems. They differ in degree of automation and provide full operational safety with the air–acetylene or nitrous oxide–acetylene flames.

The models 2280 and 2380 are microcomputer-controlled atomic absorption spectrophotometers. The use of the microcomputer greatly simplifies the operation of the instrument and makes instrument calibration a simple operation, at the same time providing versatility for the analyst.

Another advantage of the microcomputer electronics in the models 2280 and 2380 is the instrument's ability to alert the analyst to any error-producing conditions. Error codes are displayed on the readout. For example, if the

Table 2.2. *Summary of precision test results for cadmium determination* (μg l^{-1})

Laboratory	Statistical method	Sample	Spiked sample	Mean recovery % spike	LOD
1	x̄	0.234	0.729	99.0%	0.02
	sd	0.013	0.026		
	rsd %	2.8	5.5		
2	x̄	1.023	1.545	104.6%	0.02
	sd	0.044	0.060		
	rsd %	4.3	3.9		
3	x̄	0.240	2.970	98.0%	0.09
	sd	0.045★	0.049		
	rsd %	19	1.6		
4	x̄	0.084	4.818	97.3%	0.13
	sd	0.202	0.183		
	rsd %	4.3	3.9		
5	x̄	0.051	0.298	98.7%	0.09
	sd	0.037	0.090★		
	rsd %	72	30		
6	x̄	0.186	2.452	90.9%	0.11
	sd	0.088★	0.342		
	rsd %	47	14		
7	x̄	0.337	0.790	91.3%	0.01
	sd	0.026	0.030		
	rsd %	7.7	3.8		
8	x̄	0.196	4.926	94.6%	0.14
	sd	0.041★	0.116		
	rsd %	32★	21		

standards have been entered out of sequence, an error code (E-11) will be displayed on the readout. A quick referral to the instrument manual will provide the cause and remedy for the error message.

Background absorption is a problem for analysts to consider when utilizing the atomic absorption technique. Background absorption is a collective term used to describe the combined effects of such phenomena as flame absorption, molecular absorption and light scattering on atomic absorption determinations. If no correction is made for background absorption, sample absorbance may appear to be more significant than it is and the analytical result can be erroneously high. Specific applications where background correction is necessary are

1 HGA graphite furnace analyses
2 flame determinations of low concentrations of an element in the presence of high concentrations of dissolved salts
3 flame analyses where sample matrix may show molecular absorption at the wavelength of the resonance line and
4 flame determination of an element at a wavelength where flame absorption is high

Both the models 2280 and 2380 offer an optional background corrector with a deuterium arc lamp.

Some flame instruments, e.g. the Varian Spectra AA-40 has a fully automated lamp turret, providing the mounting for and automatic selection of eight lamps. When an element is selected the appropriate lamp is brought into the operating position automatically. During automated analysis, the next lamp needed is being warmed up for maximum stability.

With many flame–graphite furnace instruments now available, the flame-to-furnace changeover can be brought about in a few seconds.

Graphite tube design

The GTA-96 graphite tube atomizer used by Varian is contained in a cell fitted at each end with removable quartz windows. A stream of inert gas such as nitrogen or argon flows through the cell, protecting the graphite tube from oxidation. This gas emerges from the sampling hole carrying with it the products of drying and ashing. The graphite tube is electrically heated by current passing along its length.

When considering the purchase of a flame or graphite furnace spectrometer the following are some of the important parameters that should be considered in addition to the all important question of cost and reliability:

1 monochromator design
2 monochromator dual grating option
3 automatic wavelength drive option
4 photometer type and design
5 background correction option
6 lamp turret option
7 printer option
8 readout facilities
9 graphics option
10 burner–atomizer design

11 automation of flame gas controls
12 microcomputer facilities
13 availability of training courses
14 repair and servicing facilities

2.2.2 Autosamplers

Gilson and PS Analytical supply autosamplers suitable for automation of atomic absorption spectrophotometry and inductively coupled plasma spectrometry.

The Gilson autosampler can house up to 300 samples and is capable of operation 24 hours per day. PS Analytical supply 20- and 80-position autosamplers.

For many applications such as hydride analysis, conventional multi-element analysis and repetitive analysis for major element quantification, conventional autosamplers are insufficiently sophisticated. The PS Analytical 20.020 twenty-position autosampler has been specifically developed to fill this void. It is easily interfaced to computer systems via a TTL logic interface.

The PSA 20.020 Autosampler can operate in one of two modes: (1) as a conventional autosampler where the wash and sample items are taken from the autosampler or (2) as a slave device where both times are controlled externally. The autosampler can be restarted externally to complete a repeat analysis of the turntable set.

The PS Analytical Autosampler provides an ideal unit for automating hydride analysis by ICP, DC plasma or AA spectroscopy, and can be linked to almost any computer or instrumental system via the standard interface.

2.2.3 Applications

Cadmium in coastal waters

In Table 2.2 are presented results obtained in determinations of cadmium in eight such samples in different laboratories using a variety of methods.

With the exception of laboratories 3 and 4, which used flame atomic absorption, all laboratories participating in this study used the graphite furnace technique. Preconcentrations factors by solvent extraction of between $\times 10$ and $\times 200$ were used. It is seen that for most laboratories satisfactorily relative standard deviations ranged between 2.8 and 19 (sample) and 1.6 and 21 (spiked sample). Laboratories 5, 6 and 8 did not perform as well in this respect. Limits of detection ranged between 0.01 and $0.14\,\mu g\,l^{-1}$ cadmium and did not seem to be related to the pre-concentration factor used.

Table 2.3. *ASTM round robin QC check sample* $(\mu g\,l^{-1})$

Element	Range	Average	True	AARC
Al	62–129	86	61	65
As	19–26	23	24	24
Be	24–37	27	24	25
Cd	5.2–8.3	6.6	6.5	6.6
Cr	3.0–6.9	4.7	4.4	4.8
Co	28–37	32	30	30
Cu	2.8–11	8.4	8.7	8.5
Fe	−1.4–26	15	16	16
Pb	26–42	31	30	30
Mn	6.4–10.6	7.9	7.9	8.1
Ni	5.7–11	8.4	8.7	8.5
Se	1–9	7.3	8.7	9.0
V	50–83	70	78	76
Ba	NA	NA	NA	NA

It thus seems that, at the worst, across a number of laboratories, cadmium can be determined by the graphite furnace technique in amounts down to $0.11\,\mu g\,l^{-1}$ with a recovery of 91–105%.

Miscellaneous elements in water

In Table 2.3 are reviewed results obtained on synthetic standard solutions by the 21 laboratories involved in a round robin arranged by the American Society of Testing Materials (Schrader *et al.*, 1983).

In the final column (AARC) are results obtained using one particular instrument, namely the Varian AA975 with a GTA95 graphite furnace attachment. It is seen that this instrument gave results very close to the true values.

Toxic metals in river sediments

Flame and graphite furnace atomic absorption spectrometry have adequate sensitivity for the determination of metals in sediment samples. In this technique up to 1 g of dry sample were digested in a microwave oven for a few minutes with 5 ml aqua regia in a small PTFE lined bomb and then the bomb washings transferred to a 50 ml volumetric flask prior to analysis by flame atomic absorption spectrometry (Table 2.4). Detection limits $(mg\,kg^{-1})$ achieved by this technique were: 0.25 (cadmium, zinc); 0.5 (chromium,

Table 2.4. *Analysis 1 NBS 1645 standard sediment*

	Certif analysis mg kg^{-1}	*Certif range* mg kg^{-1}	*Aqua regia digestion* mg kg^{-1}	*Recovery aqua regia digestion* (%)
Cd	10.2 ± 1.5	8.7–11.7	8.57, 8.77 Mean 8.67	85.0
Cu	109 ± 19	90–128	104.6	96.0
Pb	714 ± 28	686–742	721	101.0
Ni	45.8 ± 2.9	42.9–48.7	46.3	101.0
Zn	1720 ± 169	1551–1889	1514, 1651 Mean 1582	92.0
Fe	113 000 ± 12 000	101 000–125 000	114, 130	101.0
Mn	785 ± 97	688–882	651, 714 Mean 682	86.9

manganese); 1 (copper, nickel and iron) and 2.5 (lead). Application of this technique to an NBS centrified sediment gave recoveries ranging between 85% (cadmium) and 101% (lead, nickel and iron) with an overall recovery of 95%.

2.3 Zeeman atomic absorption spectrometry

2.3.1 Instrumentation

Some available instrumentation for Zeeman atomic absorption spectrometry is given in Table 2.5.

The Perkin-Elmer 5000 system

The Zeeman 5000 system is a further approach to atomic absorption spectroscopy. This system offers superb features for the determination of very low elements concentrations in complex matrices with high solids contents, e.g. seawater, and provides an excellent capability for accurately correcting for very high levels of non-specific absorption (background).

Background absorption has always been the most common type of interference in electrothermal atomic absorption spectroscopy with graphite

Table 2.5. *Available Zeeman atomic absorption spectrometers*

Supplier	Model	Microprocessor	Type	Hydride and mercury attachment	Autosampler
Perkin-Elmer	Zeeman 3030	Yes (method storage on floppy disk)	Integral flame/graphite furnace	–	Yes
	Zeeman 5000	Yes, with programmer	Fully automated integral flame/graphite furnace double-beam operation roll-over protection	Yes	Yes
Varian	SpectrA A30/40	Yes, method storage on floppy disk	Automated analysis of up to 12 elements; roll-over protection	–	Yes
	SpectrA A300/400	Yes, total system control and colour graphics; 90 methods stored on floppy disk	Automated analysis of up to 12 elements; roll-over protection	–	Yes

furnace spectrometers. Most atomic absorption spectrometers for use with the graphite furnace technique are therefore equipped with continuum source background correctors which automatically compensate for broadband absorption interferences. These are sufficient for many routine applications.

Background correction with a continuum light source is, however, limited, by definition, to background absorption which is uniform within the observed spectral bandwidth. In addition background signals can be corrected only up to certain limits which are between 0.5 and 1.0 absorbance. Background absorbance higher than 1 absorbance cannot usually be fully compensated by a continuum source background corrector and may lead to erroneous results.

Non-uniform background absorption may be caused by atomic lines of matrix elements accompanying the analyte element or by a rotational fine structure of molecular spectra. Structured background absorption may lead to over- or under-compensation when using a continuum source background corrector.

With the transverse AC Zeeman system used in the Zeeman 5000 system, it becomes possible to correct easily and accurately for background absorption up to 2.0 absorbance, even when the background exhibits fine structure.

The 5000 system includes the Zeeman furnace module, the furnace programmes and the model 5000 atomic absorption spectrometer. The Zeeman module is located to the right of the model 5000 and is interfaced to the model 5000 with transfer optics.

Naturally, the model 5000 can be used for double-beam flame operation as well as with other flameless devices such as the MH 510 or the MHS-20 mercury hydride systems. Change-over from the Zeeman effect furnace operation to an alternative sampling technique can be done in seconds by simply turning a control.

Figure 2.2 illustrates the operating principle of the Zeeman 5000 system. For Zeeman operation, the source lamps are pulsed at 100 Hz (120 Hz) while the current to the magnet is modulated at 50 Hz (60 Hz). When the field is off, both analyte and background absorption are measured at the unshifted resonance line. This measurement directly compares with a 'conventional' AA measurement without background correction. However, when the field is on, only background is measured since the σ absorption line profiles are shifted away from the emission line while the static polarizer, constructed from synthetic crystalline quartz, rejects the signal from the π components. Background correction is achieved by subtraction of the field-on signal from the field-off signal. With this principle of operation, background absorption of up to 2 absorbance can be corrected most accurately even when the background shows a fine structure.

In assessing overall performance with a Zeeman effect instrument the subject of analytical range must also be considered. For most normal class transitions, σ components will be completely separated at sufficiently high magnetic fields. Consequently, the analytical curves will generally be similar to those obtained by standard AA. However, for certain anomalous transitions some overlap may occur. In these cases, curvature will be greater and may be so severe as to produce double-valued analytical curves. Figure 2.4, which shows calibration curves for copper, illustrates the reason for this behaviour. The Zeeman pattern for copper (324.8 nm) is particularly

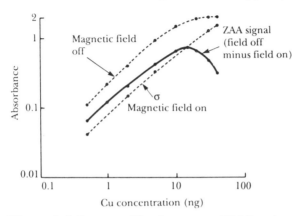

Figure 2.4 *Copper calibration curves (324.8 mm) measured with the Zeeman 5000*

complex due to the presence of hyperfine structure. The dashed lines represent the separate field-off and field-on absorbance measurements. As sample concentration increases, field-off absorbance begins to saturate as in standard AA. The σ absorbance measured with the field-on saturates at higher concentrations because of the greater separation from the emission line. When the increase in σ absorbance exceeds the incremental change in the field-off absorbance, the analytical curve, shown as the solid line, rolls over back towards the concentration axis. This behaviour can be observed with all Zeeman designs regardless of how the magnet is positioned or operated. The existence of roll-over does introduce the possibility of ambiguous results, particularly when peak area is being measured.

2.3.2 Applications

Trace metals in seawater

In Table 2.6 are presented some results for metals in a reference seawater sample NASS-1 obtained by workers at Perkin-Elmer Ltd (Grobenski *et al.* 1984). In general, the determined values are in very good agreement with the reference values even at the low concentrations concerned.

Table 2.6. *Trace metals in seawater reference material NASS – 1 STPF/Zeeman graphite furnace AAS*

Element	Certifed value ($\mu g \, l^{-1}$)	Found ($\mu g \, l^{-1}$)	Detection limit ($\mu g \, l^{-1}$)
As	1.65 ± 0.19	1.3 ± 0.5	0.2
Cd	0.29 ± 0.004	0.31 ± 0.007	0.01
Cr	0.184 ± 0.016	0.181 ± 0.020	
Mo	11.5 ± 1.9	11.7 ± 1.8	
Mn	0.022 ± 0.007	0.02 ± 0.01	0.01
Ni	0.257 ± 0.027	0.25 ± 0.06	0.05

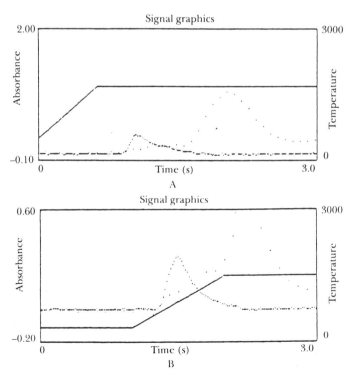

Figure 2.5 *Cadmium in sea water: (a) from the platform, note tailing of the peak; (b) from the tube wall, note symmetry of the peak*

In Figure 2.5 is shown a typical absorbance–time curve obtained in the determination of cadmium in seawater using a Varian SpectrAA 30/40 Zeeman atomic absorption spectrometer.

For this determination an aqueous solution of 2% ammonium oxalate was used as chemical modifier. An aqueous cadmium nitrate standard $(3.2\,\mu g\,Cd\,l^{-1})$ in 0.1% nitric acid was placed in the sample dispenser. The sampler was used to automatically prepare two standard additions.

The ammonium oxalate modifier separated the cadmium atomic peak from the background signal, making the use of a platform unnecessary. The polynomial interpolation of the background signal ensured accurate correction despite the large and rapid changes of the background. Argon was used as the normal gas.

In open ocean waters cadmium levels can be extremely low. To measure very low concentrations of cadmium, multiple injections of the sample may be necessary and can be achieved automatically using the programmable sample dispenser. For extremely low cadmium concentrations, extraction of the cadmium may be necessary. The APDC/MIBK (or DIBK) extraction can be used (APDC, ammonium pyrrolidine dithiocarbamate; MIBK, methyliso-butylketone; and DIBK, diisobutylketone). DIBK is the preferred solvent as it has a much higher boiling point, a low evaporation rate, is far less soluble in the aqueous layer, and the solubility of the aqueous solution in DIBK is extremely low, markedly reducing carryover.

If solvent extraction is used, the SpectrAA-40 Zeeman facilities of heated injection and variable injection rate can be used to advantage in handling the organic solvents.

Analysis of non-saline samples

Excellent precision was obtained in the determination of chromium in EPA and NBS certified standards using the Varian SpectrAA 30/4 instrument (Table 2.7).

Table 2.7.

	Value found	*Certified value*
BPA	9.75	10.2 ± 1 $\mu g\,l^{-1}$ chromium
NBS 16436	18.65	$18.6 \pm 0.4\,\mu g\,l^{-1}$ chromium

2.4 Atom-trapping technique

The sensitivity difference between direct flame and furnace atomization is now gradually being bridged via the general method of atom trapping as proposed by Watling (1977). A silica tube is suspended in the air–acetylene flame. This increases the residence time of the atoms within the tube and therefore within the measurement system. Further devices such as water-cooled systems which trap the atom population on cool surfaces and then subsequently release them by temporarily halting the coolant flow, are actively being pursued at the moment.

The application of atom-trapping atomic absorption spectrometry to the determination of lead and cadmium in potable water has been discussed by Hallam and Thompson (1986).

2.5 Vapour generation atomic absorption spectrometry

In the past certain elements, e.g. arsenic, antimony, bismuth, selenium, tellurium, mercury, germanium, tin, lead, bismuth were difficult to measure by direct atomic absorption spectrometry (Godden and Thomerson 1980; Rence 1982; Goulden and Brooksbank 1974; Stockwell 1979; Dennis and Porter 1981; Pahlavanpour *et al.* 1981). In the past 10 years or so, a novel technique of atomization, known as vapour generation via generation of the metal hydride, has been evolved, which has increased sensitivity and specificity enormously for these elements. In these methods the hydride generator is linked to an atomic absorption spectrometer (flame graphite furnace) or inductively coupled plasma optical emission spectrometer (ICP-OES) or an inductively coupled plasma mass spectrometer (ICP-MS). Some typical detection limits achievable by these techniques for the determination of elements in seawater are listed in Table 2.8.

Table 2.8.

Element	Detection limit $(\mu g \, l^{-1})$
Arsenic	3
Antimony	0.3
Bismuth	0.02
Selenium	0.09

This technique makes use of the property that the above elements exhibit, i.e. the formation of covalent, gaseous hydrides which are not stable at high temperatures. Arsenic, selenium, tellurium, antimony, bismuth and tin (and to a lesser degree lead and germanium) are volatilized by the addition of a reducing agent like sodium tetrahydroborate (III) to an acidified solution. Mercury is reduced to the atomic form in a similar manner. Other systems have used titanium (II) chloride/magnesium powder and tin (II) chloride/ potassium iodide/zinc powder as reducing agents. As a general principle sodium tetrahydroborate (III) is the preferred method because it gives faster hydride formation, higher conversion efficiency, lower blank levels and is more simple to use.

After the gaseous hydride is formed it is passed, together with any excess hydrogen gas, directly to the ICP torch, DCP system or to the atomizer mounted in the light beam of the atomic absorption spectrophotometer. Flames and tubes (which can be flame or electrically heated) have been used to atomize the elements from their hydrides. When the hydride is injected into an air–acetylene flame, extremely poor signal-to-noise (S/N) ratios occur because the resonance lines of arsenic and selenium, for instance, occur in the UV end of the spectrum. The air/acetylene flame absorbs 62% of the available light energy at 193.7 nm, the arsenic wavelength. The introduction of the argon or nitrogen/hydrogen/entrained air flame in 1967 gives much better S/N ratios, because it absorbs only 15% of the light. However, the use of the cooler flames gives rise to interference due to molecular absorption and incomplete salt dissociation.

These problems have largely been overcome by the use of heated tubes. The light beam of the spectrophotometer is focused through the tube and is never affected by the flame which lies outside the tube; hence flame absorption is eliminated. When linking to DCP the hydrides must be sheathed in argon to direct flow into the central arc of the DCP. Linking to the ICP may require a change in the automatic power control settings but is dependent on a continuous generation of the hydride and hydrogen or the plasma will be extinguished.

When the chemical composition of a material is analysed, either by atomic absorption IC P/MS or fluorescence, the sample must normally be in a liquid form. This is then aspirated or nebulized by a high-pressure gas flow. This gas, argon in the case of ICPs and other support gas such as oxygen or air in the case of atomic absorption, produces a very fine aerosol (very small liquid droplets suspended in a gas) which is carried to the 'analytical region' by the gas. In the case of ICP, the analytical region is the high-temperature (7000°C) glowing electrical discharge, and in the case of atomic absorption the analytical region is usually a flame. Unfortunately, the nebulization process is very inefficient. Typically the nebulizer converts only about 3% of the sample volume into useful aerosol. For a flow rate of 2 ml min^{-1} into the nebulizer, only 0.6 ml min^{-1} of sample actually reaches the analytical region. However, when the hydride generator is used, the nebulizer is not needed and the rate of sample presentation to the analytical region is changed. Sample typically is pumped to the hydride reaction zone at a rate of 10 ml min^{-1} in this zone. Those elements capable of forming gaseous hydrides react and these hydrides are subsequently released from the liquid phase into the argon gas flowing through the interface vessel. This argon gas, in the case of an ICP, is then directed into the plasma analytical region for analysis. The gaseous conversion is approximately 100% so that approximately $10 \, \text{ml} \, \text{min}^{-1}$ of sample are directed to the analytical region in contrast to the $0.06 \, \text{ml} \, \text{min}^{-1}$ as in the case of nebulization. This produces an increase, by a factor of 160, in the rate of sample presentation to the analytical region, which obviously increases the analytical sensitivity by a similar factor.

The vast majority of publications refer to batch devices which have been automated to a greater or lesser extent. The device described by Rence (1980) presents a state-of-the-art batch system. However, these systems are subject to one major disadvantage in that the hydrogen and hydride formed in the reaction chamber are forced into the atom cell simultaneously and disturb the equilibrium conditions, the measured signal being the sum of disturbances. In many batch systems, the apparatus has to be disassembled after one analysis. This approach can be tolerated for AA systems but has no application in ICP/OPS, ICP/MS or DCP systems.

In the batch system, sodium tetrahydroborate (III) is added to a solution of the sample and hydrochloric acid, the hydride formed is then stripped from the solution with an inert gas. Continuous-flow methods can also be used; here, the hydride is stripped in a gas–liquid separator.

The batch system gives better sensitivity but the continuous system is more suitable for automation. It gives a continuous, flame-like signal, the absorbance depending mainly upon the analyte concentration and the sample flow rate. A batch system gives a peak type signal and the absorbance maximum depends upon the analyte mass, not its concentration. The sensitivity can be increased by simply applying larger volumes of sample solution and volumes of up to $50 \, \text{ml}$ can usually be handled without problems. When large sample volumes are used the sensitivity of a batch system may be as much as an order of magnitude better than that of a continuous-flow system.

In contrast, the continuous approach relies on the continuous generation of hydrogen and a steady flow into the analyses device. Despite some objections, the inductively coupled plasma system can accommodate a reasonable level of hydrogen input without a deleterious effect on the system. Goulden and Brooksbank (1974) described an analytical system based on Technicon autoanalyser technology; however, this pumped an aluminium slurry and sulphuric acid streams – the latter to dry the hydride prior to entry into an atom cell. Stockwell (1979) described the inherent disadvantages of the Goulden system, and outlines the advantages in that developed by Dennis and Porter (1981), again using the aluminium slurry approach.

2.5.1 Instrumentation

Automating the sodium tetrahydroborate system based on continuous-flow principles represents the most reliable approach in the design of commercial instrumentation. Thompson *et al.* (1981) described a simple system for multi-element analysis using an ICP spectrometer, based on the sodium tetrahydroborate approach. PS Analytical Ltd developed a reliable and robust commercial analytical hydride generator system, along similar lines, but using different pumping principles to those discussed by Thompson *et al.* (1981). A further major advantage of this range of instruments is that different chemical procedures can be operated in the instrument with little, if any, modification. Thus, in addition to sodium tetrahydroborate as a reductant, stannous chloride can be used for the determination of mercury at very low levels.

The main advantage of hydride generation atomic absorption spectrometry for the determination of antimony, arsenic, selenium, etc. is its superior sensitivity. The limits of determination which can be routinely obtained with a batch system $(0.01-0.04\,\mu g\,l^{-1})$ and compares well with those of graphite-furnace atomic absorption spectrometry $(0.2-1\,\mu g\,l^{-1})$, another well-known technique for ultra-trace element determination. A batch system gives better sensitivity than a continuous-flow system while a quartz tube atomizer is better than a flame one.

Hydride generation techniques are not completely free from problems of interference. Interferences can arise in the liquid phase caused by changes in the hydride generation rate and/or by a decreased fraction of the analyte reduced or released from the sample solution. Gas phase interferences can occur during transport of the hydride due to a delay or loss of the analyte en route or inside the atomizer.

Continuous-flow hydride generator produced by PS Analytical Ltd

The PSA 10.002 hydride generator has, since its introduction, been widely used as a continuous source of hydride generation in ICP, DCP and most recently atomic absorption spectroscopy. Its inherent design features allow the user of each of these techniques to provide a stable baseline signal with low noise characteristics. In addition, the gas–liquid separator system provides signal rise times and also minimal memory effects. In both of these areas, which are fundamental to obtaining excellent detection levels and precision, the pumping action is critical.

Recently the PSA 10.002 has been significantly improved by the addition of a Watson Marlow 304D pump head arrangement. This avoids deterioration of pump tubes, prolongs the tube life because the tension can be released from the head and increases the new acceptability of performance by improving the methods of connecting and disconnecting the tubing itself and also the head mechanism to the motor.

For ICP applications it is important to force the hydrides into the plasma, whereas for AA applications a compromise must be reached to allow sufficient residence time to provide a measurement whilst retaining the hydrides in a stable form.

To complement this hydride generator which couples easily to all makes of AA, ICPs and DCPs, PSA has designed a complete range of mechanical interfaces. These are designed to allow the user to switch quickly between conventional analysis through a nebulizer, to the use of the hydride generator. Change-over takes less than 30 seconds, thereby removing the user resistance to making changes to hydride analysis. Autosamplers are also available. An interface system is available of suitable IBM-compatible software to operate both hydride generator and autosamplers.

Detection limits achieved for some of the above elements by the PS Analytical model 10.002 system coupled with various types of detector are between $0.15\,\mu g\,l^{-1}$ (bismuth) and $0.4\,\mu g\,l^{-1}$ (selenium).

Continuous-flow and discrete analyser to determine down to $0.05\,\mu g\,l^{-1}$ of mercury supplied by PS Analytical Ltd

Fluorescence method
The fluorescence technique utilizes the PSA 20.020 autosampler linked to the PSA 10.002 hydride generator; the vapour produced is then transferred into a PSA 10.022 fluorescence monitor. The interface between the hydride and fluorescence unit provides a flow of mercury vapour in argon, which is sheathed in an additional argon stream to reduce any quenching effects. Two argon flows are therefore provided, one to pick up mercury from the separator and the other as the sheath gas. The PSA 10.002 is used in its normal configuration, although the stannous chloride reaction scheme is preferred. In this chemical scheme there is no direct agitation produced; as with the sodium borohydride method, a modified gas–liquid mercury separator is used (PSA 10.101B). With this separator at a high rate ($2\,l\,min^{-1}$) the reagents are fed into the separator at a different level, the transfer of the mercury into the measurement cell is improved and stable readings are obtained. The mercury separator fits into the standard fixtures in the PSA 10.002.

The fluorescence system has extremely low levels of detection and is capable of rapid analysis. Typically the mercury signal will reach a peak in about 30 s and also return to the baseline in the same time. Sampling rates in the region of 40/hour with detection levels better than 0.0025 p.p.b. are obtainable at maximum sensitivity.

Atom-trapping technique
PS Analytical has recently developed a completely automatic system, the PSA 10.500, for the analysis of low levels of mercury by atomic spectroscopy. This is a combination of the hydride generator system with an amalgam-trapping facility, the PSA 10.500.

Using the PSA 10.002 hydride generator in the cold vapour mode, mercury in the samples is continuously generated as the metallic vapour. This is then trapped on a Pt/Au foil as the amalgam. The mercury can be concentrated in this way as sample is continuously pumped. On completion of the mercury timing cycle the PSA 10.002 triggers the initiation of the amalgam system. After a delay period the Pt/Au foil is rapidly heated to force off the trapped mercury as a transient pulse into the atomic cell of an atomic absorption system. Using this method the detection levels for mercury can be substantially reduced. Complete automation can be achieved using the PSA 20.020 large-volume autosampler.

The VGA-76 vapour generator accessory supplied by Varian AG

The VGA-76 vapour generation assembly is fully compatible with both the SpectrAA 10 single-beam and SpectrAA 20 double-beam atomic absorption spectrometers.

The VGA-76 system pumps the sample through a manifold where it is automatically acidified and mixed with sodium borohydride. The resulting vapour is swept by a stream of nitrogen into a quartz atomization cell (heated with an air–acetylene flame) in the atomic absorption spectrometer.

2.6 Inductively coupled plasma optical emission spectrometry

This is a further technique that has been recently introduced for the analysis of metals in waters and other materials.

An inductively coupled plasma is formed by coupling the energy from a radiofrequency (1–3 kW or 27–50 MHz) magnetic field to free electrons in a suitable gas. The magnetic field is produced by a two- or three-turn water-cooled coil and the electrons are accelerated in circular paths around the magnetic field lines that run axially through the coil. The initial electron 'seeding' is produced by a spark discharge but once the electrons reach the ionization potential of the support gas further ionization occurs and a stable plasma is formed.

The neutral particles are heated indirectly by collisions with the charged particles upon which the field acts. Macroscopically the process is equivalent to heating a conductor by a radio-frequency field, the resistance to eddy-currents flow producing Joule heating. The field does not penetrate the conductor uniformly and therefore the largest current flow is at the periphery of the plasma. This is the so-called 'skin' effect and coupled with a suitable gas-flow geometry it produces an annular or doughnut-shaped plasma. Electrically, the coil and plasma form a transformer with the plasma acting as a one-turn coil of finite resistance.

The properties of an inductively coupled plasma approach closely to those of an ideal source for the following reasons:

The source must be able to accept a reasonable input flux of the sample and it should be able to accommodate samples in the gas, liquid or solid phases.

The introduction of the sample should not radically alter the internal energy generation process or affect the coupling of energy to the source from external supplies.

The source should be operable on commonly available gases and should be available at a price that will give cost-effective analysis.

The temperature and residence time of the sample within the source should be such that all the sample material is converted to free atoms irrespective of its initial phase or chemical composition. Such a source should be suitable for atomic absorption or atomic fluorescence spectrometry.

If the source is to be used for emission spectrometry, then the temperature should be sufficient to provide efficient excitation of a majority of elements in the periodic table.

The continuum emission from the source should be of a low intensity to enable the detection and measurement of weak spectral lines superimposed upon it.

The sample should experience a uniform temperature field and the optical density of the source should be low so that a linear relationship between the spectral line intensity and the analyte concentration can be obtained over a wide concentration range.

If mass spectrometric determination of the analyte is to be incorporated, then the source must also be an efficient producer of ions.

Greenfield *et al.* (1964) were the first to recognize the analytical potential of the annular inductively coupled plasma.

Wendt and Fassel (1965) reported early experiments with a 'tear-drop' shaped inductively coupled plasma but later described the medium power $1–3\,kW$, 18 mm annular plasma now favoured in modern analytical instruments (Scott 1974).

The current generation of inductively coupled plasma emission spectrometers provide limits of detection in the range of $0.1–500\,\mu g\,l^{-1}$ of metal in solution, a substantial degree of freedom from interferences and a capability for simultaneous multi-element determination facilitated by a directly proportional response between the signal and the concentration of the analyte over a range of about five orders of magnitude.

The most common method of introducing liquid samples into the inductively coupled plasma is by using pneumatic nebulization (Thompson and Walsh 1983) in which the liquid is dispensed into a fine aerosol by the action of a high-velocity gas stream. To allow the correct penetration of the central channel of the inductively coupled plasma by the sample aerosol, an injection velocity of about $7\,m\,s^{-1}$ is required. This is achieved using a gas injection with a flow rate of about $0.5–1\,l\,min^{-1}$ through an injector tube of

1.5–2.0 mm internal diameter. Given that the normal sample uptake is 1–2 ml min^{-1} this is an insuffcient quantity of gas to produce efficient nebulization and aerosol transport. Indeed, only about 2% of the sample reaches the plasma. The fine gas jets and liquid capillaries used in inductively coupled plasma nebulizers may cause inconsistent operation and even blockage when solutions containing high levels of dissolved solids, such as sea water or particulate matter, are used. Such problems have led to the development of a new type of nebulizer, the most successful being based on a principle originally described by Babington (US Patents). In these, the liquid is pumped from a wide bore tube and thence conducted to the nebulizing orifice by a V-shaped groove (Suddendorf and Boyer 1978) or by the divergent wall of an over-expanded nozzle (Sharp 1984). Such devices handle most liquids and even slurries without difficulty.

Nebulization is inefficient and therefore not appropriate for very small liquid samples. Introducing samples into the plasma in liquid form reduces the potential sensitivity because the analyte flux is limited by the amount of solvent that the plasma will tolerate. To circumvent these problems a variety of thermal and electrothermal vaporization devices have been investigated. Two basic approaches are in use. The first involves indirect vaporization of the sample in an electrothermal vaporizer, e.g. a carbon rod or tube furnace or heated metal filament as commonly used in atomic absorption spectrometry (Gunn *et al.* 1978; Matusiewicz and Barnes 1984; Tikkanen and Niemczyk 1984). The second involves inserting the sample into the base of the inductively coupled plasma on a carbon rod or metal filament support (Salin and Horlick 1979; Salin and Szung 1984).

Atomic absorption has, for the past fifteen years, proved to be the most generally useful technique for the determination of metallic elements. As the user of atomic absorption reads the current literature, however, he finds the technique of inductively coupled plasma emission is finding favour in some applications as the method of choice for metallic determinations. Certainly the high sample throughput possible with direct reading ICP spectrometers appears to be attractive. The superior detection limits for those refractory elements that are relatively insensitive by AA, such as zirconium and tungsten, are an obvious favourable characteristic. Claims have been made as to reduction in chemical and ionization interferences with the ICP and for an extended linear analytical range. A minor but still significant factor is the elimination of the instrumental manipulations required by the use of specific source lamps for atomic absorption.

Workers at Thermo Electron Ltd, a subsidiary of Jarrell Ash, have published the following comparison of the two techniques:

1 *Detection limits* Inductively coupled plasma spectrometry has similar detection limits to flame atomic absorption for most elements but is vastly superior for those elements that require a nitrous oxide/acetylene flame in

atomic absorption spectrometry, e.g. Ba, Al, W, B, P, Ti, Zr, rare earths etc., plus inductively coupled plasma spectrometry can measure non-metals such as P, S, I, Br and high-level Cl. Inductively coupled plasma spectrometry has inferior detection limits for Zn, Pb and alkali metals, although detection limits of $2 \mu g l^{-1}$ and $25 \mu g l^{-1}$ for Zn and Pb are still acceptable. Detection limits for ICP are matrix dependent.

2 *Speed* For multi-element analysis, inductively coupled plasma spectrometry is much faster than atomic absorption spectrometry, the greater the number of elements required the greater is the advantage; for example, in water authorities a typical 13-element programme for major and trace elements takes approximately 5 minutes on a dual channel instrument.

3 *Linear dynamic range* Inductively coupled plasma spectrometry has a linear dynamic range of approximately 10^6, flame atomic absorption spectrometry is approximately 10^3 at best.

4 *Interferences* Inductively coupled plasma spectrometry has virtually no chemical interferences but may have spectral interferences in a given matrix on a given elemental wavelength. Flame atomic absorption spectrometry has many chemical interferences but few spectral (ionization) interferences.

5 *Precision* Short-term precision for flame atomic absorption spectrometry and inductively coupled plasma spectrometry are similar, 0.3–2% RSD, but in the long-term precision inductively coupled plasma spectrometry is vastly superior with precisions of better than 5% over a full 8 hour day.

6 *Overnight running* Due to the fact that the plasma is not a flame there are no combustible gases, the plasma can run unattended overnight on aqueous matrices and in fact many existing plasma users have been doing this for some years, i.e. operating a 24 hour working day instead of 8 hours. No matter how automated the atomic absorption spectrometry and the safety features present, one cannot leave this instrument unattended.

7 *Simultaneous multi-element determinations* Unlike atomic absorption spectrometry the inductively coupled plasma technique is capable of simultaneous multi-element determinations.

Both techniques have their advantages; what many water chemists are finding is that it is distinctly advantageous to have both techniques in the laboratory.

In Table 2.9 are compared detection limits claimed for atomic absorption spectrometry, graphite furnace atomic absorption spectrometry and inductively coupled plasma optical emission spectrometry.

Simultaneous versus sequential inductively coupled plasma techniques

There are two main types of inductively coupled plasma spectrometers systems.

Table 2.9. *Guide to analytical values for IL spectrometers (IL 157/357/457/451/551/951/Video 11/12/22/S11/S12 Atomic Absorption Spectrophotometers ● IL Plasma-100/-200/-300 ICP Emission Spectrometers)*

Element	Wavelength (nm) AA	Wavelength (nm) ICP	AA Lamp current (mA)	Flame AA Sensitivity² (µg l⁻¹)	Flame AA Detection limit (µg l⁻¹)	Furnace AA (IL755 CTF Atomizer) Sensitivity²	Furnace AA (µg l⁻¹)	Furnace AA Detection limit (µg l⁻¹)	ICP Detection limit (µg l⁻¹)
Aluminium (Al)[1]	309.3	396.15	8	400	25	4.0	0.04	0.01	10
Antimony (Sb)	217.6	206.83	10	200	40	8.0	0.08	0.08	40
Arsenic (As)	193.7	193.70	8	400	140³	12	0.12	0.08	30
Barium (Ba)[1]	553.5	455.40	10	150	12	4.0	0.04	0.04	0.5
Beryllium (Be)[1]	234.9	313.04	8	10	1	1.0	0.01	0.003	0.1
Bismuth (Bi)	223.1	223.06	6	200	30	4.0	0.04	0.01	35
Boron (B)[1]	249.7	249.77	15	9 000	700	–	–	–	3
Cadmium (Cd)	228.8	214.44	3	10	1	0.2	0.002	0.0002	1.5
Calcium (Ca)	422.7	393.37	7	50	2	1.0	0.01	0.01	0.2
Calcium[1]	422.7	–	7	10	1	–	–	–	–
Carbon (C)	–	193.09	–	–	–	–	–	–	40
Cerium (Ce)	–	413.77	–	–	–	–	–	–	40
Caesium (Cs)	852.1	455.53	10	150	20	–	–	–	–
Chromium (Cr)	357.9	205.55	6	40	3	4.0	0.04	0.004	3
Cobalt (Co)	240.7	238.89	8	50	4	8.0	0.08	0.008	3
Copper (Cu)	324.7	324.75	5	30	1.8	4.0	0.04	0.005	1
Dysprosium (Dy)[1]	421.2	353.17	8	600	60	–	–	–	4
Erbium (Er)[1]	400.8	337.27	8	400	40	50	0.5	0.3	3
Europium (Eu)	–	381.97	–	–	–	–	–	–	2
Gadolinium (Gd)[1]	368.4	342.25	9	13 000	2 000	1600	16	8	4
Gallium (Ga)	287.4	294.36	5	400	50	5.2	0.05	0.01	15
Germanium (Ge)[1]	265.1	209.43	5	800	50	40	0.4	0.1	20
Gold (Au)	242.8	242.80	5	100	6	5.0	0.05	0.01	10
Hafnium (Hf)[1]	307.3	339.98	10	14 000	2 000	–	–	–	5
Holmium (Ho)[1]	410.4	345.60	12	660	60	90	0.9	0.7	1

Element									
Indium (In)	303.9	325.61	5	180	30	11	0.11	0.02	15
Iridium (Ir)[1]	208.8	224.27	15	1 500	500	170	1.7	0.5	8
Iron (Fe)	248.3	238.20	8	40	5	3.0	0.03	0.01	2
Lanthanum (La)[1]	550.1	333.75	10	22 000	2 000	58	0.58	0.5	2
Lead (Pb)	217.0	220.35	5	100	9	4.0	0.04	0.007	25
Lithium (Li)	670.8	670.78	8	16	1	4.0	0.04	0.01	2[5]
Lutetium (Lu)	–	261.54	–	–	–	–	–	–	0.2
Magnesium (Mg)[1]	285.2	279.55	3	3	0.3	0.07	0.0007	0.0002	0.1
Manganese (Mn)	279.5	257.61	5	20	1.8	1.0	0.01	0.0005	1
Mercury (Hg)	253.7	253.65	3	2 500	140	40	0.4	0.2	12
Molybdenum (Mo)[1]	313.3	202.03	6	200	25	12	0.12	0.03	4
Neodymium (Nd)[1]	492.5	401.23	10	5000	700	–	–	–	8
Nickel (Ni)	232.0	221.65	10	60	5	20	0.2	0.05	4
Niobium (Nb)[1]	334.9	309.42	15	12 000	2 000	–	–	–	6
Osmium (Os)[1]	290.9	225.59	15	1 000	90	270	2.7	2	0.6
Palladium (Pd)	247.6	340.46	5	140	20	13	0.13	0.05	13
Phosphorus (P)	213.6	213.62	8	125 000	30 000	4900	49	20	16
Platinum (Pt)	265.9	214.42	10	1 000	50	80	0.8	0.2	16
Potassium (K)	766.5	766.49	7	10	1	0.4	0.004	0.004	30[5]
Praseodymium (Pr)[1]	495.1	390.84	15	20 000	2 000	–	–	–	20
Rhenium (Re)	346.1	221.43	15	8 000	800	1000	10	10	6
Rhodium (Rh)	343.5	343.49	5	200	2	20	0.20	0.1	8
Rubidium (Rb)	780.0	–	10	30	2	–	–	–	–
Ruthenium (Ru)[1]	349.9	240.27	10	800	400	–	–	–	8
Samarium (Sm)[1]	429.7	359.26	10	3 000	500	–	–	–	8
Scandium (Sc)[1]	391.2	361.38	10	100	20	–	–	–	0.5
Selenium (Se)	196.0	196.03	12	300	80[3]	8.0	0.08	0.05	30
Silicon (Si)[1]	251.6	251.61	12	800	60	60	0.60	0.6	6
Silver (Ag)	328.1	328.07	3	30	1.2	0.5	0.005	0.001	3
Sodium (Na)	589.0	589.59	8	3	0.4	0.4	0.004	0.004	7

Table 2.9. (Continued)

Element	Wavelength (nm)		AA Lamp current (mA)	Flame AA		Furnace AA (IL755 CTF Atomizer)			ICP
	AA	ICP		Sensitivity² (µgl⁻¹)	Detection limit (µgl⁻¹)	Sensitivity²	(µgl⁻¹)	Detection limit (µgl⁻¹)	Detection limit (µgl⁻¹)
Strontium (Sr)	460.7	407.77	12	80	6	1.8	0.018	0.01	0.2
Tantalum (Ta)¹	271.5	240.06	15	10 000	800	–	–	–	13
Tellurium (Te)	214.3	214.28	7	200	30	7.0	0.07	0.03	20
Terbium (Tb)¹	432.7	350.92	8	3 300	1 000	–	–	–	3
Thallium (Tl)	276.8	276.79	8	100	30	4.0	0.04	0.01	27
Thorium (Th)	–	283.73	–	–	–	–	–	–	8
Thulium (Tm)	–	313.13	–	–	–	–	–	–	0.9
Tin (Sn)¹	235.5	189.99	6	1 200	90	7.0	0.07	0.03	30
Titanium (Ti)¹	364.3	334.94	8	900	60	50	0.50	0.3	1
Tungsten (W)¹	255.1	207.91	15	5 000	500	–	–	–	14
Uranium (U)¹	358.5	263.55	15	100 000	7 000	3100	31	30	70
Vanadium (V)¹	318.5	309.31	8	600	25	40	0.40	0.1	3
Ytterbium (Yb)	398.8	328.94	5	80	–	1.3	0.01	0.01	1
Yttrium (Y)¹	410.2	371.03	6	1 800	200	1300	13	10	0.7
Zinc (Zn)	213.9	213.86	3	8	1.2³	0.3	0.003	0.001	2
Zirconium (Zr)¹	360.1	343.82	10	10 000	2 000	–	–	–	2

¹ Nitrous oxide/acetylene flame (AA).
² Sensitivity is concentration (or mass) yielding 1% absorption (0.0044 absorbance units).
³ With background correction.
⁴ Furnace AA concentration values are based on cuvette capacity of 100 µl.
⁵ Requires use of red-sensitive PMT.

Monochromator system for sequential scanning

This consists of a high-speed, high-resolution scanning monochromator
which views one element wavelength at a time. Typical layouts are shown in
Figure 2.6. Figure 2.6(a) represents a one-channel air path double
monochromator design with a pre-monochromator for order sorting and stray
light rejection and a main monochromator to provide resolution of up to
0.02 nm. The air-path design is capable of measuring wavelengths in the
range of 190–900 nm. The wide wavelength range enables measurements to

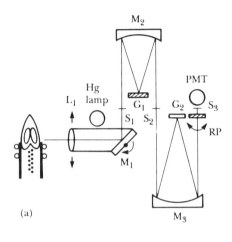

(a)

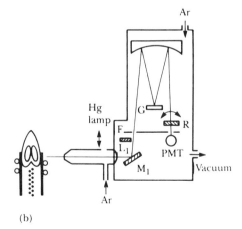

(b)

Figure 2.6 *(a) A double monochromator consisting of an air-path monochromator
with a pre-monochromator for order sorting and stray light rejection to determine
elements in the 190–900 nm range; (b) the vacuum UV monochromator – an
evacuated and argon-purged monochromator to routinely determine elements in the
160 to 500 µm range*

be performed in the ultraviolet, visible and near infrared regions of the spectrum (allowing determinations of elements from arsenic at 193/70 nm to caesium at 852.1 nm).

The second channel (Figure 2.6(b)) is a vacuum monochromator design allowing measurements in the 160–500 nm wavelength range. The exceptionally low wavelength range gives the capability of determining trace levels of non-metals such as bromine at 163.34 nm as well as metals at low UV wavelengths such as the extremely sensitive aluminium emission line at 167.08 nm. Elements such as sulphur, boron or phosphorus can be routinely determined using interference-free emission lines.

The sequential instrument, equipped with either or both monochromators facilitates the sequential determination in a sample of up to 63 elements in turn at a speed as fast as 18 elements per minute. Having completed the analysis of the first sample, usually in less than a minute, it proceeds to the second sample, etc.

Polychromator system for simultaneous scanning

The polychromator systems scan many wavelengths simultaneously, i.e. several elements are determined simultaneously at higher speeds than are possible with monochromator systems. It then moves on to the next sample. A typical system is shown in Figure 2.7.

It is possible to obtain instruments which are equipped for both sequential and simultaneous scanning, e.g. the Labtam 8410.

For the water chemist the decision whether to purchase a sequential or a simultaneous system depends on obtaining answers to the following questions:

1 What matrices do you want to analyse? The more diverse the matrices the greater the need for a sequential instrument. If the matrix is very well defined, e.g. wear metals in oils, the case for a simultaneous system becomes stronger.
2 How many elements are required and could this number increase? The simultaneous plasma is custom built so a decision has to be made at time of purchase on how many analytical channels are put on the focal curve and also on the best wavelengths for the defined matrices and concentration levels expected.
3 How many samples need to be analysed per day? The great advantage of simultaneous inductively coupled plasma is speed. In many cases full analysis can be done within 1 minute, whether it is 1 element or 61 elements. With sequential inductively coupled plasma the larger the number of elements, the longer the analysis time.
4 What are the relevant advantages of either technique to achieve accurate results? Simultaneous inductively coupled plasma has fixed positions by which background correction can be applied leading to possible errors on

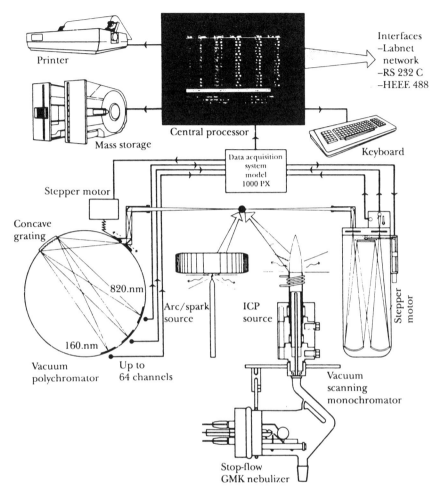

Figure 2.7 *Polychromator system for inductively coupled plasma atomic emission spectrometer*

some elements in some matrices. Inter-element corrections are routinely required on simultaneous inductively coupled plasma but a bad wavelength for a given element in a given matrix may not be possible even if inter-element correction is applied. A sequential inductively coupled plasma can be programmed to avoid most or all of any encountered spectral interferences.

5 What are the relevant precisions of the two inductively coupled plasma versions? Sitting on a peak (i.e. sequential) will always give better precision and reproducibility than scanning past the peak (i.e. simultaneous) (as most scanning inductively coupled plasma systems do). Low-level

detection limits may be superior on the simultaneous inductively coupled plasma in terms of reproducibility over a period of time.

Hybrid inductively coupled plasma techniques

Chromatography-inductively coupled plasma

Direct introduction of the sample into an inductively coupled plasma produces information only on the total element content. It is now recognized that information on the form of the element present, or trace element speciation, is important in a variety of applications including water analysis. One way of obtaining quantitative measurement of trace element speciation is to couple the separation power of chromatography to the inductively coupled plasma as a detector. Since the majority of interesting trace metal speciation problems concern either involatile or thermally unstable species, e.g. ions such as tributyltin, methyl mercury or trimethyllead, HPLC becomes the separation method of choice. The use of HPLC as the separation technique requires the introduction of a liquid sample into the inductively coupled plasma with the attendant sample introduction problem. An example of this is the separation of four arsenic species, arsenate, monomethyl arsonic acid, arsenite and monomethyl arsinic acid on a reverse phase C 18 column using dilute sulphuric acid as the eluent. The eluent is then merged with a flow of sodium borohydride solution in a continuous flow hydride generator, and the liberated hydrides swept into the inductively coupled plasma. This approach has been used to determine arsenic species in estuarine, soil pore and drinking waters.

Flow injection with inductively coupled plasma

In conventional inductively coupled plasma optical emission spectroscopy, a steady-state signal is obtained when a solution of an element is nebulized into the plasma. In flow injection (Ruzicka and Hansen 1978) a carrier stream of solvent is fed through a 1 mm i.d. tube to the nebulizer continuously using a peristaltic pump, and into this stream is injected, by means of a sampling valve, a discrete volume of a solution of the element of interest. When the sample volume injected is suitably small a transient signal is obtained (as opposed to a steady-state signal which is obtained with larger sample volumes) and it is this transient signal that is measured. Very little sample dispersion occurs under these conditions, the procedure is very reproducible and sampling rates of 180 samples per hour are feasible.

Inductively coupled plasma with atomic fluorescence spectrometry

Atomic fluorescence is the process of radiational activation followed by radiational deactivation, unlike atomic emission which depends on the collisional excitation of the spectral transition. For this, the inductively coupled plasma is used to produce a population of atoms in the ground state and a light source is required to provide excitation of the spectral transitions. Whereas a multitude of spectral lines from all the accompanying elements are emitted by the atomic emission process, the fluorescence spectrum is relatively simple, being confined principally to the resonance lines of the element used in the excitation source.

The inductively coupled plasma (ICP) is a highly effective line source with a low background continuum. It is optically thin – it obeys Beer's law – and therefore exhibits little self-absorption. It is also a very good atomizer and the long tail flame issuing from the plasma has such a range of temperatures that conditions favourable to the production of atoms in the ground state for most elements are attainable. It is therefore possible to use two plasmas in one system, a source plasma to supply the radiation to activate the ground rate atoms in another, the atomizer. This atomic fluorescence (AFS) mode of detection is relatively free from spectral interference, the main drawback of inductively coupled plasma optical emission spectroscopy.

Good results have been obtained using a high power (6 kW) ICP as a source and a low power (<1 kW) plasma as an atomizer.

2.6.1 Instrumentation

The chemist interested in purchasing an inductively coupled plasma optical emission spectrometer is faced with an embarrassingly large variety of suppliers (see Table 2.10). The eventual choice will be made on the basis of selecting a sequential or a simultaneous instrument for general requirements (e.g. degree of automation and computerization) and cost (price range £40 000–£90 000).

The Perkin-Elmer ICP 5000 combined graphite furnace spectrometer inductively coupled plasma emission spectrometer is a fairly sophisticated instrument with the following features:

- Every instrumental parameter is controlled by the built-in microprocessor or the external data system
- High-energy monochromator has both an 84 × 84 mm holographic grating blazed at 210 nm and a ruled grating blazed at 580 nm, covering a spectral range of 170 to 900 nm in the first order.
- Storage and recall of six complete AA programs by the microprocessor
- Correction of non-specific atomic absorption over the complete wavelength range available for the AA mode
- Optical interface:
  - Transmits either ICP emission or radiation from the AA source into the monochromator
  - Allows change from AA to ICP in two seconds with the turn of a knob
  - Both optical interface and monochromator can be purged for working in the far UV at wavelengths down to 170 nm

Table 2.10. *Inductively coupled plasma optical emission spectrometers available on the market*

Supplier	Model	System	Number of elements claimed	Maximum analysis rate (elements min^{-1})	Microprocessor	Autosampler	Range (nm)
Perkin-Elmer	Plasma II	Optimized sequential system	70	Up to 50	Yes	Yes	160–800
Perkin-Elmer	ICP 5500	Sequential		15	Yes	Yes	170–900
Perkin-Elmer	ICP 5500B	Sequential		20	Yes	Yes	170–900
Perkin-Elmer	ICP 6500	Sequential		20	Yes	–	170–900
Perkin-Elmer	ICP 5000	Can be used for flame and graphite furnace ASS and inductively coupled plasma atomic emission spectrometry (sequential)		–	Yes	Yes	175–900
Perkin-Elmer	Plasma 40	Lower-cost sequential			Personal computer control	Yes	160–800
Labtam	Plasma scan 8440	Simultaneous (polychromator or with optional monochromator for sequential)	60–70	Up to 64	Yes	Yes	170–820

Labtam	Plasma 8410	Sequential	More than 70	–	Yes	Yes	170–820
Thermoelectron	Plasma 300 (replacing the Plasma 200) (single (air) or double (air/ vacuum) options available	Sequential	Up to 63	Up to 18 (single channel air); up to 24 (double channel air/ vacuum)	Yes	Yes	160–900
Philips	PV 8050 series PV 8055 PV 8060 PV 8065	Simultaneous	56	–	Yes	Yes	165–485 and 530–860
Philips	PU 7450	Sequential	70	–	Yes	Yes	190–800
Baird	Spectrovac PS3/4 plasma hydride device option	Simultaneous and sequential	Up to 60	Up to 80 samples per hour each up to 60 elements	Yes	Yes	162–766 and 162–800
Baird	Plasmatest system 75	Simultaneous and sequential	Up to 64	–	Yes	Yes	175–768 and 168–800
Spectro Analytical Ltd	Spectroflame plasma hydride device option	Simultaneous and sequential	Up to 64	–	Yes	Yes	165–800

- Provides the adjustment of viewing height of the ICP torch ICP emission source and power supply:
- High analytical reproducibility by automatic tuning of the plasma to maintain its most efficient operating level
- Power adjustable to 2500 watts for analysis of all types of aqueous and organic sample solutions
- Sensors for argon and water flow prevent accidental damage to the torch components

The first enthusiastic reports on analytical applications of ICP gave the impression that the ultimate technique for the determination of metallic and semimetallic elements had been discovered. However, as time passed it became clear that while ICP emission substantially enhances the capability of atomic spectroscopy as an analytical technique ICP must be considered a complement to, rather than a replacement for, atomic absorption. And indeed, each technique offers advantages and disadvantages of its own, as described below.

ICP emission

- Excellent detection limits for refractory elements that are relatively insensitive to determination by flame AA including U, B, P, Ta, Ti and W. They can be determined at submilligram/litre levels with ICP.
- Multi-element analysis is more rapid by ICP if more than eight elements are to be determined in the same sample
- Matrix dependence is less of a problem with ICP which is practically free of ionization and chemical interferences
- Although ICP does not offer the intrinsic specificity of atomic absorption, the spectral interferences which are observed may be eliminated by background correction techniques or the use of alternate wavelengths

Flame atomic absorption

- The analytical precision of AA is usually better than 0.3% for many analyses and cannot currently be matched by ICP. This is particularly important for applications where high precision is required.
- The sample throughput of flame AA is higher for single-element determinations of the automated determination of up to six to eight elements per sample. This is due to the rapid equilibration of the AA signal after sampling begins.

Graphite furnace atomic absorption

- The HGA graphite furnace offers outstanding detection limits and is the method of choice for trace analysis of the majority of elements.

- If the sample quantity is limited, the HGA graphite furnace should be selected for its microsampling capability. Microlitre quantities of sample are sufficient to perform an analysis.
- Direct analysis of solid samples (e.g. sediments, plant materials) is possible only with the HGA graphite furnace technique.

Sequential ICP analysis

The sequential ICP multi-element determines each element under its own optimum, uncompromised set of analytical conditions. By necessity, simultaneous multi-element instruments always operate under compromised conditions. However, sequential multi-element instruments can be programmed to perform each elemental analysis under specific optimized conditions.

Some detection limits achieved by the inductively coupled plasma atomic emission systems are given in Table 2.9.

Autosampler for ICP

The Labtam autosampler has been specifically designed for inductively coupled plasma applications and provides total flexibility for both simultaneous and sequential operations. The design and construction employed in this autosampler has simplified difficult and arduous applications commonly experienced in laboratories with high sample throughput.

2.6.2 Applications

Determination of cadmium and lead

These measurements were carried out on a Thermo-electron Plasma 300 ICP optical emission spectrograph with a vacuum monochromator run under the following conditions:

- Plasma 300 dual channel ICP spectrometer
- Power level 1 (950 W)
- Torch height 3 mm nebulizer pressure 30 lb in^{-2}
- Pb 220.35 with background correction on LHS
- Cd 214.44 with background correction on LHS

The results obtained indicate that both lead and cadmium meet the detection limits necessary to meet EEC legislation for determinations of lead (50 μg l^{-1}) and cadmium (5 μg l^{-1}) in potable water.

Table 2.11. *Detection limits obtained: Thermo-electron Plasma 300 sequential inductively coupled plasma atomic emission spectrometer*

Element	Detection limit $(\mu g l^{-1})$	Element	Detection limit $(\mu g l^{-1})$
Na	60.6	Cd	2.8
K	513	Pb	60.6
Mg	14	Cr	7.9
Ca	23.3	Mn	1.16
Al	23	Fe	6.5
Cu	6.5	Ni	13.5
Zn	2.8		

13-element analysis in water

Table 2.11 gives detection limits achieved in the analysis of 13 commonly occurring elements using a dual channel air plus vacuum monochromator in a Thermoelectron Plasma 300 sequential inductively coupled plasma atomic emission spectrometer. The sample was preconcentrated 10-fold by evaporation.

A power lead of 950 W was used with a peristaltic pump speed of $1 \, ml \, min^{-1}$ with a 30 s pump delay. Better detection limits were obtained than by atomic absorption spectrometry in most cases. The complete analysis took 6 minutes.

Analysis of acid rain

The analysis of acid rain has become an important political and economic subject due to the damage done to forestry areas, lakes and buildings. Although aluminium is of primary importance, ten other elements give important data.

Unlike river/potable waters the calcium and magnesium levels are very low, i.e. $1–5 \, mg \, l^{-1}$, so the more sensitive lines can be utilized and because of the low calcium content the primary aluminium line can be used to lower the detectable levels of aluminium.

No spectral problems are encountered and due to the low levels required primary lines are used in most cases. In scanning ICP a peak search must be performed to ensure good routine analysis. One problem that arises for very low levels is that the 'peak' may be so small as not to be detected. An increase in integration time will increase the photon count to improve the detectability, but at the expense of increasing analysis time.

In Table 2.12 is shown the satisfactory results obtained applying the Baird Spectrovan system to National Bureau of Standards water samples of known composition.

Table 2.12. *Determination of trace metals in water (SRM 1643a) using ICP-AES with ultrasonic nebulization*

Element	Measured value[1] (μg l^{-1})	NBS value (μg l^{-1})
Ag	4.7 ± 1.8	2.8 ± 0.3
As	69.9 ± 2.4	76 ± 7
Ba	43.1 ± 2.0	46 ± 2
Be	17.9 ± 1.4	19 ± 2
Cd	11.7 ± 1.1	10 ± 1
Co	19.4 ± 0.9	19 ± 2
Cr	16.3 ± 1.1	17 ± 2
Cu	18.2 ± 2.2	18 ± 2
Fe	92.6 ± 4.4	88 ± 4
Mn	28.7 ± 0.9	31 ± 2
Mo	94.0 ± 6.1	95 ± 6
Ni	49.7 ± 2.7	55 ± 3
Pb	25.8 ± 1.4	27 ± 1
Se	12.8 ± 1.4	11 ± 1
Zn	68.3 ± 2.8	72 ± 4

[1] Ten consecutive measurements.

2.7 Inductively coupled plasma optical emission spectrometry – mass spectrometry

Inductively coupled mass spectrometry combines the established inductively coupled plasma to break the sample into a stream of positively charged ions which are subsequently analysed on the basis of their mass. Inductively coupled mass spectrometry does not depend on indirect measurements of the physical properties of the sample. The elemental concentrations are measured directly – individual atoms are counted giving the key attribute of high sensitivity. The technique has the additional benefit of unambiguous spectra and the ability to measure different isotopes of the same element directly.

The sample under investigation is introduced, most typically in solution, into the inductively coupled plasma at atmospheric pressure and a temperature of approximately 6000 K. The sample components are rapidly dissociated and ionized and the resulting atomic ions are introduced via a carefully designed interface into a high-performance quadrupole mass spectrometer at high vacuum.

A horizontally mounted ICP torch forms the basis of the ion source. Sample introduction is via a conventional nebulizer optimized for general-purpose solution analysis and suitable for use with both aqueous and organic solvents.

Nebulized samples enter the central channel of the plasma as a finely dispersed mist which is rapidly vaporized; dissociation is virtually complete during passage through the plasma core with most elements fully ionized.

Ions are extracted from the plasma through a water-cooled sampling aperture and a molecular beam is formed in the first vacuum stage and passes into the high-vacuum stages of the quadrupole mass analyser.

In an inductively coupled plasma mass spectrometry system a compact quadrupole mass analyser selects ions on the basis of their mass-to-charge ratio, m/e.

The quadrupole is a simple compact form of mass analyser which relies on a time-dependent electric field to filter the ions according to their mass-to-charge ratio.

Ions are transmitted sequentially in order of their m/e with constant resolution, across the entire mass range.

A multi-stage vacuum system provides the required low pressure in the analyser and is under computer control, enabling the control of the transition from atmospheric pressure at the sampling cone to the 2×10^{-4} torr within the quadrupole to be invisible to the operator.

In addition to the need for good resolution between peaks and high sensitivity, another instrument performance parameter, abundance sensitivity, is of particular importance to this technique.

Abundance sensitivity is a measure of the ability of the spectrometer to detect a small peak in the presence of a very large adjacent peak. Often the requirement is for detection limits of nanograms per litre or better in the presence of major elements at nearby masses (e.g. ultratrace levels of toxic impurities in the presence of large peaks of matrix elements).

A variety of sample introduction techniques which have been used successfully with ICP optical emission instruments depend on the measurement of transient signals. For a quadrupole with a scanning detector it is necessary to scan significantly faster than the analyte signal is changing and many discrete sample introduction systems have pulses which last a few seconds. With the plasma quad instrument manufactured by VG Isotopes Ltd, for example, it is possible to perform a sweep over all the elements in 100 ms.

ICP-MS has some significant advantages over ICP-OES when looking at pulses of this kind, such as greater sensitivity, simpler spectra and wider element coverage.

This technique is being adopted cautiously by several water laboratories. Its considerable cost has certainly been a deterrent to all but the most enterprising laboratories. It does seem to satisfy the seemingly conflicting demands made of chemists in water-control laboratories for high sample throughput and extremely low detection limits.

Automatic, multi-element operation, with semiquantitative analysis of the entire mass range in less than a minute (with subsequent quantitation by comparison with standards) and detection limits at least an order of magnitude better than any alternative techniques are most impressive credentials.

For many elements the most useful spectral lines arise from singly charged ions of which the inductively coupled plasma is an abundant source. These ions can be extracted for mass spectrometry. Gray (1975) pioneered the use of plasmas fed by solutions and realized the potential of the ICP as a source for MS. He was joined at Surrey University by Date and together with Fassel's group at Ames, Iowa, they established ICP-MS as a viable analytical technique (Gray 1975; Houk 1980; Date and Gray 1981, 1983; Gray and Date 1983).

In principle, the inductively coupled plasma offers many advantages as an ion source for mass spectrometry. Making it work, however, presents some formidable difficulties. The ICP reaches temperatures of 5000–9000 K at its core, operates with $10–20 \, \text{l} \, \text{min}^{-1}$ of argon support gas and, of course, operates at atmospheric pressure. The mass spectrometer, on the other hand, operates at 10^{-5} torr or below and obviously cannot be subjected to the heat of the plasma. Clearly, then, the critical aspect in the development of ICP-MS is in the design of a suitable interface between the two.

The interface consists essentially of two orifices; the first, known as the 'sampler' is a water-cooled metal cone containing a circular hole of 0.5–1.0 mm diam. A second metal cone, known as a 'skimmer', is placed behind the sampler and divides an initial expansion chamber (~1 torr) from the vacuum region that contains the ion optics and quadrupole mass filter. Expansion through the sampler orifice produces a supersonic beam of ions and neutral argon atoms which is further narrowed by the exclusion of peripheral ions at the succeeding skimmer nozzle. The sampler is immersed in the so-called 'normal analytical region' of the plasma, 5–15 mm above the RF load coils used for plasma generation. Because the sampler is able to accept gases from the axial channel of the plasma over a cross-section of around two to three times the orifice diameter a sizeable fraction of the central plasma channel is sampled.

Complementary aspects of inductively coupled plasma emission spectrometry and mass spectrometry

The combination of a mass spectrometer with an inductively coupled plasma has proved to be a harmonious relationship enabling the speed and convenience of sample introduction into the inductively coupled plasma to be combined with the high sensitivity and isotope-ratio capability of mass spectrometry. Some complementary features of the ICP-MS union are summarized in Table 2.13, demonstrating that many of the weaknesses of existing sample-introduction techniques for the inorganic mass spectrometer can be overcome when an inductively coupled plasma is used as an ion source.

Figures 2.8(a) and (b) compare the conventional optical spectrum for a solution containing cerium with a mass spectrum for the same element. The experimental facilities and operating conditions used in obtaining these spectra are summarized on page 49.

Table 2.13. *Complementary aspects of the ICP-MS*

ICP emission spectrometry	*Mass spectrometry*
Efficient but mild ionization source (produces mainly singly charged ions)	Ion source required
Sample introduction for solutions is rapid and convenient	Sample introduction can be difficult for inorganic samples. Thermal spark-source, or secondary ion sources are generally restricted to solid samples and are time-consuming
Sample introduction is at atmospheric pressure	Often requires reduced-pressure sample introduction
Few matrix or interelement effects are observed and relatively large amounts of dissolved solids can be tolerated	Limited to small quantities of sample
Complicated spectra with frequent spectral overlaps	Relatively simple spectra
Detectability is limited by relatively high background continuum over much of the useful wavelength range	Very low background level throughout a large section of the mass range
Moderate sensitivity	Excellent sensitivity
Isotope ratios cannot usually be determined	Isotope-ratio determinations are possible

ICP-MS system

The facilities in the ICP-MS experiments consisted of an argon–ICP system incorporating a 2.5 kW, 27.12 Mi-iz (Plasma Therm model HFP-2500F, Kresson, N,H.) plasma power supply and impedance-matching unit with a conventional-sized, water-cooled three-turn load coil and a plasma torch of optimized low-flow, low-power design. A laboratory-constructed concentric pneumatic nebulizer with a Scott type double-pass spray chamber was used for sample introduction. In the spectra the plasma was operated at 1.25 kW with a nebulizer flow of $0.8 \, \mathrm{l \, min^{-1}}$ and a coolant flow of $12 \, \mathrm{l \, min^{-1}}$. The mass-spectral system consisted of an interface arrangement, a series of ion lenses to focus the ion beam, a quadrupole mass spectrometer (QMG 511) Balzers, Hudson, N,H.) and a detector (secondary electron multiplier) located off the optical axis to prevent entry of extraneous photons.

ICP-OPS system

The ICP system in this case was operated with a 2.5 kW, 40.68 MHz power supply (Plasma Therm, model HFL-2000L) at 1.0 kW with a nebulizer flow of $1.0 \, \mathrm{l \, min^{-1}}$ and a coolant flow of $13 \, \mathrm{l \, min^{-1}}$, but was otherwise similar to that described above. An optical detection scheme consisting of a Heath monochromator (model EU-700) and an RCA 1P28 photomultiplier was used to record the spectrum.

The optical spectrum contains dozens of strong lines and hundreds of weaker lines for cerium, all of which are superimposed on top of an already complicated background emission spectrum resulting from emission lines of the argon-support gas and molecular bands from atmospheric contaminants (such as OH, NH, N_2). In addition, a relatively large continuous background is observed (Figure 2.8(a)). The corresponding mass spectrum (Figure 2.8(b)) on the other hand, is considerably simpler and contains fewer background peaks especially above $m/z = 40$. The response due to cerium appears at the m/z positions for cerium isotopes (mass units 140 and 142) respectively. A substantial improvement in spectral simplicity and discrimination against background response is readily apparent when mass spectral detection is employed.

Mass spectral interferences are still apparent, however, as the major background ions from Ar^+, ArH^+ and O_2^+ (Figure 2.8(b)) interfere with the major isotopes of calcium and sulphur ($^{40}Ca^+$ and $^{32}S^+$) requiring that minor isotopes be substituted for these elements. Other interferences are less severe, for example, minor ions ArO^+ and Ar_2^+ may interfere with iron and selenium respectively. Molecular ions derived from reagent acids, such as ArN^+, ClO^+, $ArCl^+$, SO^+ and SO_2^+ must be taken into account in sample preparation. The background spectra for the reagent acids nitric acid, sulphuric acid and hydrochloric acid have recently been studied in detail,

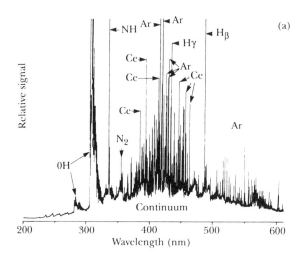

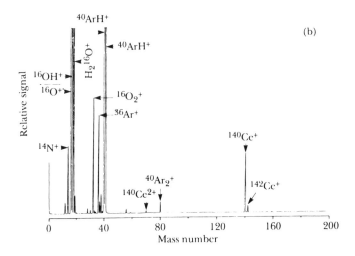

Figure 2.8 *Comparison between the optical emission detection (a) for 100 p.p.m. Ce in the ICP and mass spectral detection; (b) for 10 p.p.m. of the same element. The optical spectrum contains dozens of strong Ce lines in the 350–450 nm region in addition to a number of strong background features due to OH, N₂, NH, etc. (only the most intense features are labelled). The corresponding mass spectrum is simpler and shows clearly resolved Ce isotopes at 140 and 142 mass units. (The minor isotopes at 136 and 138 are not visible at the detection sensitivity shown.) A small peak due to Ce²⁺ is also apparent at m/e = 70*

sulphuric and hydrochloric acids give complex spectra, making nitric acid the acid matrix of choice in most applications. Significantly, no major background is observed above $m/z = 82$. Many of the molecular ions present in the background appear to result from clustering reactions in the extraction process.

Inductively coupled plasma mass spectrometry also shows an advantage over flame atomic absorption spectrometry which has essentially similar detection limits to those for inductively coupled plasma – optical emission spectrometry. Comparison between inductively coupled plasma, mass spectrometry and electrothermal atomic absorption spectrometry is more difficult because inductively coupled plasma mass spectrometry responds to the concentration of analyte present in a solution while electrothermal atomic absorption spectrometry responds to the absolute mass of analyte deposited upon the atomizer. Nevertheless, if a sample volume of 20 µl is assumed, then electrothermal atomic absorption spectrometry detection limits are in the range of $0.005-5 \mu g l^{-1}$. Inductively coupled plasma mass spectrometry is therefore competitive with electrothermal atomic absorption spectrometry for ultra low-level analysis, but offers significantly greater convenience and speed and, in addition, the capability for multi-element determinations which are not possible in conventional electrothermal atomic absorption spectrometry.

In inductively coupled plasma mass spectrometry, the linear dynamic range, based on ion counting, for most elements is five decades, i.e. similar to inductively coupled plasma optical emission spectrometry.

A major advantage of inductively coupled plasma mass spectrometry is the unique capacity for isotope ratio determination. Precision in these measurements is usually in the range 0.1–1% RSD.

Reports on interference effects in inductively coupled plasma mass spectrometry are mixed. Many workers report relative freedom from inter-element effects similar to that which had been found previously for inductively coupled plasma optical emission spectrometry. Little suppression of ionization occurs in the presence of $1000 \, mg l^{-1}$ sodium (Gray and Date 1983; Gray 1986) the classical interference of phosphate on calcium determinations is absent (Douglas *et al.* 1983).

Conversely, other workers (Horlick *et al.* 1986; Pickford and Brown 1986) report severe and widespread inter-element effects in inductively coupled plasma mass spectrometry.

To sum up, the inductively coupled plasma mass spectrometry method offers remarkably high detection powers with the capability for isotope-ratio determination with speed and convenience. However, at this present stage of development a number of limitations are apparent: samples with highly dissolved solids cannot be optimally analysed unless they are diluted to at least 0.2% of total dissolved solute; instrument drift can be severe (5–10% per hour); interference effects appear to be more severe than in inductively coupled plasma optical emission spectrometry, accuracy of analysis is generally worse than with inductively coupled plasma optical mass spectrometry and capital costs are high (approximately US $2 000 000 for commercial instruments). Strenuous efforts are currently being made to minimize these limitations.

2.7.1 Applications

Inductively coupled plasma mass spectrometry has been applied to the analysis of waters and marine sediments (Douglas and Quan 1983; Yorkshire Water Authority).

Workers at the Yorkshire Water Authority have evaluated the VG Isotopes Ltd plasmaquad inductively coupled plasma mass spectrometer during 1984 and 1986. The instrument failed to meet a specification they had laid down for its use in water analysis. Failure was on both reliability and performance grounds. They then went on to study sequential inductively coupled plasma optical emission spectrometry and indeed, based on their experience in these trials, ultimately purchased two of these instruments (see section on inductively coupled plasma optical emission spectrometry).

2.7.2 Instrumentation

Several manufacturers, including VG Isotopes and Perkin-Elmer can now supply equipment for inductively coupled plasma mass spectrometry.

The Perkin-Elmer Elan 500 instrument is discussed below. This spectrometer is designed for routine and rapid multi-element quantitative determinations of trace and ultratrace elements and isotopes. The Elan 500 can determine nearly all of the elements in the periodic table with exceptional sensitivity.

The entire Elan 500 Plasmalok system is designed for simplicity of operation. A typical daily start-up sequence from the standby mode includes turning on the plasma and changing to the operating mode. After a brief warm-up period for the plasma, routine sample analysis can begin.

Plasmalok eliminates undesirable secondary discharges in the interface region. The result is increased sampling orifice lifetime and minimal contamination from vaporized orifice materials. Plus, a sharply reduced presence of doubly charged ions simplifies the mass spectrum.

More importantly, Plasmalok helps maintain ion energies at constant low levels with minimal spreads, even when plasma conditions are changed. Even though the Elan 500's optics (the electrical devices that focus the stream of ions) are easily adjusted, Plasmalok ensures that such adjustments are rarely required for routine analysis.

An exclusive feature of the Perkin-Elmer Elan 500 instrument is the Omnirange system. The high sensitivity of inductively coupled plasma mass spectrometry makes it an ideal technique for trace and ultratrace analysis. However, the sensitivity can be a hindrance when determining sample components of widely diverse ion concentrations. Omnirange allows the analyst to selectively reduce the sensitivity for specific individual analyte masses directly from the system computer. Normally, an analyst would have to bring signals within the working range by sample dilution. With Omnirange, sample dilutions can be avoided. The user can bring off-scale peak mass values into the useful working range simply by indicating which mass values are to have reduced sensitivity. Omnirange extends the useful upper limits of the technique without sacrificing sensitivity and performance.

V.G. Isotopes Ltd are another leading manufacturer of inductively coupled plasma mass spectrometers. The special features of their VG Plasmaquad PQ2 include a multi-channel analyser which ensures rapid data acquisition over the whole mass range. The multi-channel analyser facilities include 4096 channels, 300 m facility for spectral analysis, user-definable number of measurements per peak in peak jumping mode and the ability to monitor data as it is acquired. A multi-channel analyser is imperative to acquire short-lived signals from accessories such as flow injection, electrothermal vaporization, laser ablation, etc. or for fast multi-element survey scans (typically 1 minute).

A variant of the Plasmaquad PQ2 is the Plasmaquad PQ2 plus instrument. This latter instrument has improved detector technology which incorporates a multimode system which can measure higher concentrations of elements without compromising the inherent sensitivity of the instrument. This extended dynamic range system (Table 2.14) produces an improvement in effective linear dynamic range to eight orders of magnitude. Hence, traces at microgram per litre levels can be measured in the same analytical sequence as major constituents.

This technique has been applied to the analysis of river waters, sludges, effluents and brines and seawaters containing up to 2% solids.

2.8 Inorganic mass spectrometry

Finnigan MAT manufacture a range of mass spectrometric equipment for isotope ratios and inorganic elemental analysis.

Model 251 is an exceptionally precise, fully automated stable isotope ratio mass spectrometer for gaseous samples. It achieves performance standards far beyond those generally available for analysis of H/D, $^{13}C/^{12}C$, $^{15}N/^{14}N$, $^{18}O/^{17}O/^{16}O$ and $^{34}S/^{32}S$ and noble gases.

Model 281 is a high-precision, fully computer-controlled UF_6 mass spectrometer for the simultaneous determination of uranium isotope

Table 2.14. *Dynamic ranges of various techniques*

Graphite furnace AAS	$0.1\,\mu g l^{-1}$ to $1\,mg l^{-1}$
Plasmaquad PQ2	$0.1\,\mu g l^{-1}$ to $10\,mg l^{-1}$
Inductively coupled plasma–atomic absorption spectrometry	$10.0\,\mu g l^{-1}$ to $1000\,mg l^{-1}$
Plasmaquad PQ2 plus	$0.1\,\mu g l^{-1}$ to $1000\,mg l^{-1}$

abundance, including the ^{235}U and ^{238}U minor isotopes. These measurements are important in uranium enrichment research, production control and safeguard management of nuclear material.

Models 271 and 271/45 provide reliable, hydrocarbon-group type quantitative gas mixture and trace gas analyses with long-term stability, sensitivity and linear response. The 271 and 271/45 incorporate specially designed heated sample inlet systems for accurate and reproducible analysis and multiple resolving power for trace analysis down to the sub-parts per million level.

The Thermionic Quadrupole (THQ^{TM}) mass spectrometer provides isotope ratio determination and trace-element analysis by stable isotope dilution. It allows the analysis of small samples with high sensitivity and precision, yet at a very low price.

Delta and delta E are low cost, stable isotope ratio mass spectrometers that provide precise determination of isotope ratios of carbon, hydrogen, nitrogen, oxygen and sulphur, with automatic operation. The delta's low sample consumption and its analytical reproducibility are optimized for cost-efficient routine applications.

Delta S is an isotope ratio mass spectrometer designed specifically to accept samples directly from a gas chromatograph or CHN analyser. The unique modular inlet design provides unmatched flexibility. Optimized differential pumping allows for direct, capillary gas flow to the ion source. ISODA T^{TM} software allows data to be acquired and evaluated as the instrument is set up for the next experiment.

2.9 Visible spectrophotometry

Visible spectrophotometers are commonly used in the water industry for the estimation of colour in a sample or for the estimation of coloured products produced by reacting a colourless component of the sample with a reagent with which produces a colour, which can be evaluated spectrophotometrically.

Formerly, visible spectrophotometry was used extensively for the determination of metals. Thus, lead was determined by reaction of the lead ions with dithizone, with which it produces a red coloured complex. However, following the introduction of techniques such as atomic absorption spectrometry, these applications are becoming fewer.

Visible spectrophotometry still finds extensive use in the determination of some anions such as phosphate, silicate, chloride and nitrate, although these analyses are now normally accomplished using an autoanalyser in which the spectrophotometer is embodied (discussed elsewhere).

An extensive modern application of visible spectrophotometry is in the determination of organic substances in water including non-ionic detergents, alcohols (by estimation of the orange-red product produced upon reaction with ceric ammonium nitrate, aldehydes (by formation of the red coloured 2:4 dinitrophenyl-hydrazones) to name but a few determinations.

Some commercially available instruments, in addition to visible spectrophotometry, can also perform measurements in the UV and near IR regions of the spectrum. These have not yet found extensive applications in the field of water analysis.

Suppliers of visible spectrophotometers are reviewed in Table 2.15.

For many water industry applications a simple basic UV/visible single-beam instrument such as the Phillips PU 8620 or the Cecil Instrument CE 2343 or CE2303 or CE2202 will suffice. If better instrument stability is required, then a double-beam instrument is preferable and can be purchased at little extra cost.

Moving up-market, spectrophotometers are available which have computer interfaces which enable reaction kinetics studies to be carried out (e.g. the Cecil CE 2202, CE 2272, CE 2292, CE 2303, CE 2373 and CE 2393 single-beam instruments and the Cecil CE 594 double-beam instruments). These instruments also have an autosampler facility capable of handling up to 40 samples. Even more sophisticated is the Cecil 6000 double-beam instrument equipped with a real-time graphics plotter. In this instrument sets of parameters, i.e. methods, can be stored and curve fitting carried out. It also enables multi-component analysis to be carried out. Analyses of mixtures of up to nine different materials may be carried out.

2.10 X-ray fluorescence spectrometry

This technique has a true multi-element analysis capability and requires no foreknowledge of the elements present in the sample. As such it is very useful for the examination of many of the types of samples encountered in the water

Table 2.15. *Visible–ultraviolet–near infrared spectrophotometers*

Spectral region	Range (nm)	Manufacturer	Model	Single or double beam	Cost range
UV/visible	–	Philips	PU 8620 (optional PU 8620 scanner)	Single	Low
Visible	325–900	Celcil Instruments	CE 2343 Optical Flowcell	Single	Low
Visible	280–900	Celcil Instruments	CE 2393 (grating, digital)	Single	High
Visible	280–900	Celcil Instruments	CE 2303 (grating, non-digital)	Single	Low
Visible	280–900	Celcil Instruments	CE 2373 (grating, linear)	Single	High
UV/visible	190–900	Celcil Instruments	CE 2292 (digital)	Single	High
UV/visible	190–900	Celcil Instruments	CE 2202 (non-digital)	Single	Low
UV/visible	190–900	Celcil Instruments	CE 2272 (linear)	Single	High
UV/visible	200–750	Celcil Instruments	CE 594 (microcomputer controlled)	Double	High
UV/visible	190–800	Celcil Instruments	CE 6000 (with CE 6606 graphic plotter option)	Double	High
UV/visible	190–800	Celcil Instruments	5000 series (computerized and data station)	Double	High

UV/visible	–	Philips	PU 8800	Double	High
UV/visible	–	Kontron	Unikon 860 (computerized with screen)	Double	High
UV/visible	–	Kontron	Unikon 930 (computerized with screen)	Double	High
UV/visible	190–1100	Perkin-Elmer	Lambda 2 (microcomputer electronics screen)	Double	High
UV/visible	190–750 or 190–900	Perkin-Elmer	Lambda 3 (microcomputer electronics)	Double	Low to High
UV/visible	190–900	Perkin-Elmer	Lambda 5 and	Double	
UV/visible		Perkin-Elmer	Lambda 7 (computerized with screen)	Double	High
UV/visible	185–900 & 400–3200	Perkin-Elmer	Lambda 9 (computerized with screen)	UV/vis/NIR	High
UV/visible	190–900	Perkin-Elmer	Lambda Array 3840 (computerized with screen)	Photodiode	High

laboratory, such as deposits obtained from sewage and trade effluents, river and oceanic sediments, plant material (e.g. mosses) and animal material (e.g. crustacea). As will be discussed later, it has also been used for the examination of liquid water samples.

2.10.1 Energy-dispersive X-ray fluorescence spectrometry

Energy-dispersive X-ray fluorescence (EDXRF) spectrometry is an instrumental analytical technique for non-destructive multi-elemental analysis. The use of modern-day technologies coupled with the intrinsic simplicity of X-ray fluorescence spectra (as compared for instance with optical emission (OE) spectra) means that the powerful EDXRF technique can be used routinely in the modern water analytical laboratory. The EDXRF spectrum for iron is a clearly resolved doublet whilst the optical emission spectrum contains more than 4000 lines. This simplicity is a direct consequence of the fact that XRF spectra are a result of inner shell electron transitions which are possible only between a limited number of energy levels and for relatively few electrons. Optical emission spectra, on the other hand, arise from electron transitions in the outer, valence shells which are closer together in energy, more populated than the inner shells and from which it is easier to promote electron transitions.

In order to generate X-ray spectra, we may excite the elements in the specimen with any one of the following:

- X-ray photons
- high-energy electrons
- high-energy charged particles
- gamma rays
- synchrotron radiation.

The term XRF is generally applied when X-ray photons are used to generate characteristic X-rays from the elements in the specimen. The most commonly used sources of such X-rays (in the 2-100 keV range) are radioisotopes and X-ray tubes. An EDXRF spectrometer such as the XR300 uses a compact, low power (10–100 W typical) X-ray tube capable of delivery of X-ray photons with a maximum energy of 30 or 50 keV.

Why is the technique referred to as 'energy-dispersive' XRF?

The classical XRF spectrometer which has been commercially available since the 1950s uses crystal structures to separate (resolve) the X-rays emanating from the fluorescence process in the irradiated specimen. These crystals diffract the characteristic X-rays from the elements in the specimen, allowing them to be separated and measured. The characteristic fluorescent X-rays are said to have been separated from each other by the process of 'wavelength dispersion' (WDXRF). Each element emits characteristic lines which can be separated by WDXRF before being individually counted. For each line and diffracting crystal, we can set a detector at a particular angle (from the Bragg equation) and collect X-rays, which are primarily from the selected element.

The EDXRF system uses the Si(Li) (lithium-drifted silicon) detector to simultaneously collect all X-ray energies emitted from the specimen. Each detected X-ray photon gives rise to a signal from the detector. The magnitude of this signal is proportional to the energy of the detected X-ray and when amplified and digitized can be passed to a multi-channel analyser which displays a histogram of number of X-rays (intensity) against energy. The incident photons, therefore, have been electronically separated (dispersed) according to their energy. The energy of each of the X-rays from all the elements is readily accessible from published tables.

Due to the simple spectra and the extensive element range (sodium upwards in the periodic table) which can be covered using the Si(Li) detector and a 50 kV X-ray tube, EDXRF spectrometry is perhaps unparalleled for its qualitative element analysis power.

Qualitative analysis is greatly simplified by the presence of few peaks which occur in predictable positions and by the use of tabulated element/line markers which are routinely available from the computer-based analyser.

To date, the most successful method of combined background correction and peak deconvolution is to use the method of digital filtering and least squares (FLS) fitting of reference peaks to the unknown spectrum (Stathan 1977). This method is robust, simple to automate and is applicable to any sample type.

The combination of the digital filtering and least squares peak deconvolution method and empirical correction procedures has application throughout elemental analysis. This approach is suitable for specimens of all physical types and is used in a wide selection of industrial applications.

2.10.2 Total reflection X-ray fluorescence spectrometry

The major disadvantage of conventional energy dispersive X-ray fluorescence spectrometry has been poor elemental sensitivity, a consequence of high background noise levels resulting mainly from instrumental geometries and sample matrix effects. Total reflection X-ray fluorescence (TXRF) is a relatively new multi-element technique with the potential to be an impressive analytical tool for trace-elemental determinations for a variety of sample types. The fundamental advantage of TXRF is its capability to detect elements in the picogram (PG) range in comparison to the nanogram (NG) levels typically achieved by traditional energy-dispersive X-ray fluorescence spectrometry.

The problem in detecting atoms at the $ng\,l^{-1}$ or sub-$\mu g\,l^{-1}$ level is basically one of being able to obtain a signal which can be clearly distinguished from the background. The detection limit being given typically as the signal which is equivalent to three times the standard deviation of the background counts for a given unit of time. In energy-dispersive X-ray fluorescence spectrometry the background is essentially caused by interactions of radiation with matter resulting from an intense flux of elastic and Compton scattered photons. The background especially in the low-energy region (0–20 keV) is due in the main to Compton scattering of high-energy Bremsstrahlung photons from the detector crystal itself. In addition, impurities on the specimen support will contribute to the background. The Auger effect does not contribute to an increased background, as the emitted electrons, of different but low energy, are absorbed either in the beryllium foil of the detector entrance windows or in the air path of the spectrometer.

A reduction in the spectral background can be effectively achieved by X-ray total reflection at the surface of a smooth reflector material such as quartz. X-ray total reflection occurs when an X-ray beam impinges on a surface at less than the critical angle of total reflection. If a collimated X-ray beam impinges onto the surface of a plane smooth and polished reflector at an angle less than the critical angle, then total reflection occurs. In this case the angle of incidence is equal to the angle of reflection and the intensities of the incident and totally reflected beams should be equal.

The principles of TXRF were first reported by Yoneda and Horiuchi (1971) and further developed by Aiginger and Wodbrauschek (1974). In TXRF the exciting primary X-ray beam impinges upon the specimen prepared as a thin film on an optically flat support at angles of incidence in the region of 2 to 5 minutes of arc below the critical angle. In practice the primary radiation does not (effectively) enter the surface of the support but skims the surface, irradiating any sample placed on the support surface. The scattered radiation from the sample support is virtually eliminated, thereby drastically reducing the background noise. A further advantage of the TXRF system, resulting from the new geometry used, is that the solid-state energy-dispersive detector can be accommodated very close to the sample (0.3 mm), which allows a large solid angle of fluorescent X-ray collection, thus enhancing a signal sensitivity and enabling the analysis to be carried out in air at atmospheric pressure.

The sample support or reflector is a 3 cm diameter wafer made of synthetic quartz or perspex. The water sample can be placed directly onto the surface. The simplest way to prepare liquid samples is to pipette volumes between 1 and 50 μl directly onto a quartz reflector and allow them to dry. For aqueous solutions the reflector can be made hydrophobic (e.g. by silicon treatment) in order to hold the sample in the centre of the plate. Suitable elements for calibration can be achieved by a simple standard addition technique.

Since Yoneda and Horiuchi (1971) first reported the use of TXRF various versions have been developed (Knoth and Schwenke 1978, 1980; Schwenke and Knoth 1982; Pella and Dobbyn 1988). Recently an X-ray generator with a fine focus tube and multiple reflection optics has been developed by Seifert & Co. and coupled with an energy-dispersive spectrometer fitted with an Si(Li) detector and multi-channel analyser supplied by Link Analytical. The new system, which will be described later, known as the EXTRA II, represents the first commercially available TXRF instrument.

An example of the detection limits achieved by the Link Analytical EXTRA III (3σ above background, counting time 1000 s) are shown in Figure 2.9 for the molybdenum anode X-ray tube and for excitation with the filtered Bremsstrahlung spectrum from a tungsten X-ray tube.

The data shown was obtained from diluted aqueous solutions which can be considered to be virtually free from any matrix effects. A detection limit of 10 pg for the 10 µl sample corresponds to a concentration of $1\,\mu g\,l^{-1}$. A linear

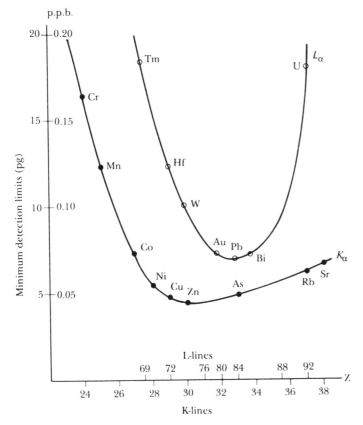

Figure 2.9 *Minimum detection limits of the TXRF spectrometer (Mo-tube 13 mA/60 kV; counting time 1000 s)*

dynamic range of four orders of magnitude is obtained for most elements: for example, lead at concentrations 2–20 000 $\mu g\,l^{-1}$ using cobalt as an internal standard at 2000 $\mu g\,l^{-1}$.

The attractive features of TXRF can be summarized as follows:

- An inherent universal calibration curve is obtained as a smooth function of atomic number
- The use of internal single-element standardization eliminates the need for matrix-matched external standards
- Only small sample volumes are required (5–50 μl)
- The technique requires only a simple sample preparation methodology.

The attractive features of the TXRF technique outlined above suggest that TXRF has the potential to become a very powerful analytical tool for trace-elemental determinations applicable to a wide range of matrix types and may, indeed, compete with the inductively coupled plasma mass spectrometry.

Instrumentation in energy and total reflection dispersive X-ray fluorescence spectrometry

Various suppliers of instruments are listed in Table 2.16.

Philips PW 1404 energy-dispersive X-ray fluorescence spectrometers

The Philips PW 1404 is a powerful versatile sequential X-ray spectrometer system developed from the PW 1400 series and incorporating many additional hardware and software features that further extend its performance.

All system functions are controlled by powerful microprocessor electronics, which make routine analysis a simple, push-button exercise and provide extensive safeguards against operator error. The microprocessor also contains sufficient analytical software to permit stand-alone emergency operation, plus a range of self-diagnostic service-testing routines.

The main characteristics of this instrument are as follows:

- Identifies all elements from boron to uranium
- Choice of side window X-ray tubes allows optimum excitation for all applications
- New detectors and crystals bring improved light element performance
- 100 kV programmable excitation enhances heavy element detection
- Special calibration features give more accurate results:
  - Auxiliary collimator provides high resolution
  - Programmable channel mask reduces background
  - Fast digital scanning speeds data collection
  - High angular accuracy aids positive identification

Table 2.16. *Energy dispersive and total reflection X-ray fluorescence spectrometers*

Supplier	Model	Type	Sample types	Computer	Handling of peak overlaps	Element range	Detector
Link Analytical	XR 200/300	Energy dispersive	Solid and liquid	Yes	Filtered least squares technique	Atomic numbers 15–55 (Mo–Ba)	10 mm^2 155 eV resolution Si (Li) detector
Pye Instruments (Philips)	PW 1404	Energy dispersive sequential	Solids and liquids	Yes	Various techniques	B–U	Argon flow scintillation
Seifert	Extra III	Energy dispersive and multiple total reflection	Solids and liquids	Yes	No correction for matrix effects required except for those in the range sodium to phosphorus	All elements beyond sodium	80–mm^2 165 eV Si (Li) detector

- Powerful software includes automatic peak labelling
- Compact one-cabinet system
- Distributed intelligence via five microprocessors
- High-frequency generator cuts running costs and improves stability
- New high-speed electronics allow operation at 1 million counts/s
- System self-selects analytical programs for unknowns
- Surface down sample presentation aids accurate analysis of liquids
- Small airlock speeds sample throughput, cuts helium costs
- Designed for laboratory automation
- Extended microprocessor control ensures simple, error-free operation
- Front panel continuously displays system status
- New generation software for DEC computers
- Computer dialogue in English, French, German, Spanish
- Colour graphics simplify results interpretation
- Extensive programming, reporting, editing facilities available

The layout of the Philips PW 1404 instrument is shown in Figure 2.10.

Unique among XRF instruments the Extra II TXRF spectrometer yields lower limits of detection in the region of 10 pg (1 pg $= 10^{-12}$ g) for more than

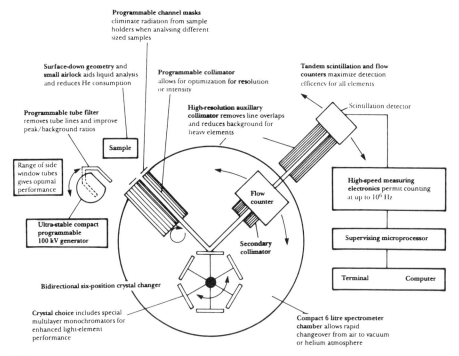

Figure 2.10 *Layout of Philips PW 1404 energy-dispersive X-ray fluorescence spectrometer*

Table 2.17. *Detection limits Seifert Extra II X-ray spectrometer*

	Detection limit (pg)
Atomic numbers 18–38 (argon to strontium) and 53–57 (iodine–lanthanum) and 78–83 (osmium to bismuth)	<5
Chlorine atomic numbers 39–43 (yttrium to technetium), 47–52 (silver to tellurium), 65–71 (terbium to lutecium) and 90–92 (thorium to uranium)	5–10
Phosphorus (15) Sulphur (16) Ruthenium (44) Rhodium (45) Palladium (46) Neptunium (93) Plutonium (94)	10–30
Aluminium (13) Silicon (14)	30–100
Sodium (11) Magnesium (12)	>100

60 elements (Table 2.17). All elements upwards from sodium ($Z = 11$) in the periodic table may be determined. The inclusion of twin excitation sources, which may be switched electronically within a few seconds, assures optimum sensitivity for all detectable elements. The applicable concentration range is from per cent to below $1 \mu g\,l^{-1}$. As little as $1 \mu g$ of sample is sufficient to determine elements at the milligram per litre level; calibration is necessary only once and is carried out during installation. The calibration will remain unchanged for a period of at least 12 months. Quantitative analysis is simple and uses the method of internal standardization. No external standards are necessary. The method requires no correction of matrix effects for all elements except those in the range sodium to phosphorus. Empirical absorption-enhancement correction models may be applied to these light elements. Sample preparation for solutions and dispersions is very simple, requiring only a micropipette. Complete digestion of materials is not mandatory. Finely divided powders may be analysed providing they are homogeneous.

Application of X-ray fluorescence spectrometry: energy-dispersive X-ray fluorescence spectrometry

Ellis and Leyden (private communication) used dithiocarbamate precipitation methods to determine between $2\,mg\,l^{-1}$ and $2\,\mu g\,l^{-1}$ of five elements in natural water samples. In Table 2.18 is shown the excellent agreement generally obtained between X-ray fluorescence results obtained using the Link XR 200/300 instrument and atomic absorption spectrometric techniques in analysis of the preconcentrates. Agreement does not extend over the whole concentration range examined for manganese. Some disparity also occurs in zinc determinations and it is believed here that the error is in the graphite furnace results.

Table 2.18. *The analysis of natural waters by dibenzyldithiocarbamate precipitation – energy-dispersive X-ray spectrometry and electrothermal atomization atomic absorption spectrometry*

Sample	Mn		Fe		Ni		Cu		Zn	
	XRF	AA	XRF	AA	XRF	AA	XRF	AA	XRF	AA
1	96	104	22	22	4	37	10	4	76	17
2	195	158	75	38	18	23	5	6	44	30
3	114	182	59	56	0.9	2.3	nd	nd	7	nd
4	450	450	46	42	0.9	2.2	1.8	5	61	51
5	2400	2700	15	22	20	25	5	14	1059	342
6	360	335	76	62	12	2.4	6	6	104	92

Total reflection X-ray fluorescence spectrometry

The Seifert Extra II total reflection X-ray fluorescence spectrometer has been applied to the determination of a range of elements in rainwater (Stossel and Prange 1985), water cycle (rain, river and sea-water sediments, particulate matter and mussel tissue) samples (Prange *et al.* *1987*) and seawater (Prange *et al.* 1985). Detection limits between 5 and $20\,ng\,l^{-1}$ were achieved for most elements. In Table 2.19 are compared results obtained on rainwater by direct analysis and concentration by freeze drying and pre-concentration using dibenzyldithiocarbamate.

Prange *et al.* (1987) determined dissolved heavy metal traces in sea water by a procedure based on total reflection X-ray fluorescence spectrometry. The trace elements are separated from the salt matrix by chelation with

Table 2.19. *Results of three rainwater samples from Pellworm obtained by using the different analytical techniques*

	Sample 1 (week 40, 1984) TXRF, n = 3							
	Direct		Freeze-drying		Reverse phase		DPASV	
Element	Mean	RSD	Mean	RSD	Mean	RSD	Mean	RSD
S	789	2.9						
K	39.3	10.3						
Ca	161.4	4.1						
Ti			<0.1					
V			0.48	8.9	0.35	9.9		
Cr			<0.1					
Mn	1.96	10.8	1.91	3.1	1.78	7.7		
Fe	6.70	4.7	6.46	7.1	5.70	3.1		
Co	$-^1$	$-^1$	$-^1$	$-^1$	$-^1$	$-^1$	0.12	8.0
Ni	0.71	20.1	0.68	7.3	0.47	3.5	0.66	7.0
Cu	0.67	5.6	0.60	2.5	0.58	3.1	(1.43	17.7)
Zn	9.49	2.3	9.49	0.9	9.16	1.3	7.33	10.5
As			0.24	7.0				
Pb	3.31	3.0	2.81	0.8	2.83	0.8	2.57	14.4
Se			0.11	1.9	0.07	8.7	(0.05	15.0)
Rb			0.09	6.5				
Sr	1.04	5.5	0.94	8.8				
Mo	0.07	30.0	0.06	17.7	0.05	12.5		
Cd			0.13	15.0	0.12	6.5	0.12	9.9
Ba			1.55	4.9				

[1] Internal standard.

sodium dibenzyldithiocarbamate, selective chromatographic adsorption of the metal complexes onto a lipophilized silica gel carrier, and subsequent elution of the metal chelates by a chloroform–methanol mixture. Aliquots of the eluate are then dispensed onto highly polished quartz sample carriers and evaporated to thin films for the TXRF measurements. The following elements can be determined: V, Mn, Fe, Co, Ni, Cu, Zn, Se, Mo, Cd, (Hg), Pb and U. For 200 ml samples and a measuring time of 1000 s detection limits of 5–20 ng kg^{-1} are achieved for most of these elements.

Some typical results are given in Table 2.20.

Table 2.20. *TXRF results compared to the sea-water reference sample NASS-1*

Element	Method used for certification[1]	Concentration (μg kg^{-1}) reference values	TXRF $n = 8$
V	–	–	1.42 ± 0.14
Cr	g, m	0.184 ± 0.016	–
Mn	i, s	0.022 ± 0.007	0.023 ± 0.005
Fe	i, m, s	0.192 ± 0.036	0.235 ± 0.020
Co	i, m	0.004 ± 0.001	0.005 ± 0.002
Ni	e, i, m, s	0.257 ± 0.027	0.230 ± 0.010
Cu	a, c, i, m, s	0.099 ± 0.010	0.092 ± 0.005
Zn	a, i, m, s	0.159 ± 0.028	0.125 ± 0.020
As	a, h	1.65 ± 0.19	–
Mo	d, m	11.5 ± 1.9	10.5 ± 0.5
Cd	a, c, i, m, s	0.029 ± 0.004	0.038 ± 0.009
Hg	–	–	0.072 ± 0.011
Pb	a, i, m	0.039 ± 0.006	0.042 ± 0.006
U	–	–	3.16 ± 0.10

[1] a – Anodic stripping voltammetry; c – Coprecipitation separation/graphite-furnace atomic absorption spectrometry (GFAAS); d – Direct determination by GFAAS; e – Chelation/hanging drop mercury electrode; g – Isotope dilution gas chromatography mass spectrometry. h – Hydride-generation AAS; i – Immobilized ligand separation GFAAS; m – Isotope dilution solid-source mass spectrometry; s – Chelation/liquid liquid extraction GFAAS.

2.11 Neutron activation analysis

This is a very sensitive technique which, amongst many other types of samples, has been applied to the analysis of waters, sediments and sewage sludges and trade effluents.

Due to the complexity and cost of the technique no water laboratory in the UK has its own facilities for carrying out neutron activation analysis. Instead, samples are sent to one of the organizations that possess the facilities, e.g. the Atomic Energy Research Establishment at Harwell or the Joint Manchester–Liverpool University Reactor located at Risley.

As mentioned above, the technique is extremely sensitive and tends to be used when a referee analysis is required on a material which then becomes a standard for checking out other methods. Another advantage of the technique is that a foreknowledge of the elements present is not essential. It can be used to indicate the presence and concentration of entirely unexpected elements, even when present at very low concentrations.

In neutron activation analysis, the sample in a suitable container, often a pure polyethylene tube, is bombarded with slow neutrons for a fixed time together with standards. Transmutations convert analyte elements into radioactive elements, which are either different elements or isotopes of the original analyte.

After removal from the reactor the product is subject to various counting techniques and various forms of spectrometry to identify the elements present and their concentration.

2.12 Polarography and voltammetry

A high proportion of trace metal analysis carried out in water laboratories is based on the techniques of atomic absorption spectrometry and inductively coupled plasma atomic emission spectrometry. Both of these methods give estimates of the total metal concentration of metal present and do not distinguish between different valency states of the same metal. Yet in many instances, and to an increasingly greater extent, a knowledge of the particular ionic species present is of great importance. To name but three examples, differences in the toxicities of the first and last named of the following three pairs is of great significance: CrIII/CrVI, CuI/CuII, and AsIII/AsV.

Polarography, unlike the techniques mentioned earlier, is capable of such speciation studies and, as such, is becoming of increasing interest to the water chemist.

Three basic techniques of polarography are of interest and the basic principles of these are outlined below:

Universal: differential pulse (DPN, DPI, DPR)

In this technique a voltage pulse is superimposed on the voltage ramp during the last 40 ms of controlled drop growth with the standard dropping mercury electrode; the drop surface is then constant. The pulse amplitude can be preselected. The current is measured by integration over a 20 ms period immediately before the start of the pulse and again for 20 ms as the pulse nears completion. The difference between the two current integrals $(l_2 - l_1)$ is recorded and this gives a peak-shaped curve. If the pulse amplitude is increased, the peak current value is raised but the peak is broadened at the same time.

Classical: direct current (DCT)

In this direct current method, integration is performed over the last 20 ms of the controlled drop growth (Tast procedure): during this time, the drop surface is constant in the case of the dropping mercury electrode. The resulting polarogram is step-shaped. Compared with classical DC polarography according to Heyrovsky, i.e. with the free-dropping mercury electrode, the DCT method offers great advantages; considerably shorter analysis times, no disturbance due to current oscillations; simpler evaluation and larger diffusion-controlled limiting current.

Rapid: square wave (SQW)

Five square-wave oscillations of frequency around 125 Hx are superimposed on the voltage ramp during the last 40 ms of controlled drop growth – with the dropping mercury electrode the drop surface is then constant. The oscillation amplitude can be preselected. Measurements are performed in the second, third and fourth square-wave oscillation; the current is integrated over 2 ms at the end of the first and the end of the second half of each oscillation. The three differences of the six integrals ($l_1 - l_2, l_3 - l_4, l_5 - l_6$) are averaged arithmetically and recorded as one current value. The resulting polarogram is peak shaped.

2.12.1 Instrumentation

Metrohm are leading suppliers of polarographic equipment. They supply three main pieces of equipment: the Metrohm 646 VA Processor, the 647 VA Stand (for single determinations) and the 675 VA sample changer for a series of determinations). Some features of the 646 VA processor are listed below:

- Optimized data acquisition and data processing
- High-grade electronics for a better signal-to-noise ratio
- Automatic curve evaluation as well as automated standard addition for greater accuracy and smaller standard deviation
- Large, non-volatile methods memory for the library of fully developed analytical procedures
- Connection of the 675 VA sample changer for greater sample throughput
- Connection of an electronic balance
- Simple, perfectly clear operation principle via guidance in the dialogue mode yet at the same time high application flexibility thanks to the visual display and alphanumeric keyboard
- Complete and convenient result recording with built-in thermal recorder/ printer

The 675 VA sample changer is controlled by the 646 VA Processor on which the user enters the few control commands necessary. The 646 VA processor also controls the 677 drive unit and the 683 pumps. With these auxiliary units, the instrument combination becomes a polarographic analysis station which can be used to carry out on-line measurements.

The 646 VA processor is conceived as a central, compact component for automated polarographic and voltammetric systems. Thus, two independent 647 VA stands or a 675 VA sample changer can be added. Up to 4 multi-dosimats of the 665 type for automated standard additions and/or addition of auxiliary solutions can be connected to each of these wet-chemical workstations. Connection of an electronic balance for direct transfer of data is also possible.

Program-controlled automatic switching and mixing of these three electrode configurations during a single analysis via software commands occur. The complete electrode is pneumatically controlled. A hermetically sealed mercury reservoir of only a few millilitres suffices for approximately 200 000 drops. The mercury drops are small and stable, consequently there is a good signal-to-noise ratio. Mercury comes into contact only with the purest inert gas and plastic free of metal traces. Filling is seldom required and very simple to carry out. The system uses glass capillaries which can be exchanged simply and rapidly.

Up to 30 complete analytical methods (including all detailed information and instructions) can be filed in a non-volatile memory and called up. Consequently, a large, extensive and correspondingly efficient library of analytical methods can be built up, comprehensive enough to carry out all routine determinations conveniently via call-up of a stored method.

The standard addition method (SAM) is the procedure generally employed to calculate the analyte content from the signal of the sample solution; electric current SAM amount of substance/mass concentration. The SAM is coupled directly to the determination of the sample solution so that all factors which influence the measurement remain constant. There is no doubt that the SAM provides results that have proved to be accurate and precise in virtually every case.

The addition of standard solutions can be performed several times if need be (multiple standard addition) to raise the level of quality of the results still further.

Normally, a real sample solution contains the substances to be analysed in widely different concentrations. Thus a large amount of zinc and copper might be accompanied by minor quantities of cadmium and lead. In a single multi-element analysis, however, all four heavy metals must be determined simultaneously. The superiority of the facilities offered by segmented data acquisition in this respect is clear when a comparison is made with previous solutions. The analytical conditions were inevitably a compromise: no matter what type of analytical conditions were selected, such large differences could rarely be reconciled. In the recording, either the peaks of zinc or copper or just those of cadmium and lead were shown meaningfully – each of the other two were either no longer recognizable or led to gigantic signals with cut-off

Table 2.21. *Polarographic differential pulse direct current square wave and anodic scanning*

Supplier	Type	Model No.	Detection limits
Metrohm	Differential Pulse Direct current Square wave	646 VA processor 647 VA stand 675 sample changer 665 Dosimat (motor driven piston burettes for standard additions)	0.05 µg l⁻¹ quoted for basic metals 2–10 µg l⁻¹ for nitriloacetic acid
	Direct current normal pulse differential pulse 1st harmonic ac. 2nd harmonic ac. Kalousek	506 Polarecord	
	Direct current sampled differential pulse	DC 626 Polarecord	

Chemtronics Ltd	On-line voltammetric analyser for metals in effluents and field work	PDV 2000	$\sim$0.1 µg l^{-1}
RDT Analytical Ltd	Differential pulse anodic stripping on-line voltammetric analyser for metals in effluents and field work	ECP 100 plus ECP 104 programmes ECP 140 PDV 200	–
	On-line voltametric analyser for continuous measurement of metals in effluents and water	OVA 2000	–
EDT Analytical Ltd	Cyclic voltammetry differential pulse voltammetry linear scam voltammetry, square-wave voltammetry, single- and double-step chronopotentiometry and chronocoulometry	Cipress Model CYSY -1B (basic system) CY57-1H-(high-sensitivity system)	

peak tips. And all too often the differences were still too large even within the two concentration ranges. Since the recorder sensitivity and also all other instrument and electrode functions could only be set and adjusted for a single substance, even automatic range switching of the recorder was of very little use.

The dilemma is solved with the 646 VA processor: the freedom to divide the voltage sweep into substance-specific segments and to adjust all conditions individually and independently of one another within these segments opens up quite new and to date unknown analytical possibilities. Furthermore, it allows optimum evaluation of the experimental data.

Various suppliers of polarographs are summarized in Table 2.21.

Metrohm, in addition to the 646 VA processor and 647 VA stand or 675 VA sample changer, which can carry out differential pulse direct current and square-wave measurements, also supply two other instruments capable of carrying out different kinds of measurements.

The SO6 Polarecord
direct current
normal pulse
differential pulse
1st harmonic a.c.
2nd harmonic a.c.
Kalousek

The 626 Polarecord
direct current sampled d.c.
differential pulse

The latter is a basic instrument intended for routine analysis and teaching applications. It does not have sensitivity of the 646 VA but has, nevertheless, been used for the determination of down to $200 \, \mu g \, l^{-1}$ levels of metals in tap water. The stripping differential pulse voltammetry at a hanging mercury drop electrode in a pH 4.7 ammonium citrate buffered medium.

2.12.2 Applications

Polarography is an excellent method for trace and ultra-trace analysis of inorganic and organic substances and compounds. The basic process of electron transfer at an electrode is a fundamental electrochemical principle and for this very reason polarography can be used as an analytical method over a wide range of applications. Its powerful detection ability – often down to the p.p.t. region with heavy metals – its high accuracy and high precision as well as its multi-element detection combine to make polarography the method of choice in numerous analytical tasks.

Determination of trace metals in water

After previous enrichment at the ranging mercury drop electrode (HMDE) seven heavy metals were all determined using differential pulse-stripping voltammetry: the first three metals via anodic dissolution (DPASV) and four other metals via cathodic dissolution (DPCSW) of the corresponding metal complexes. The determination limits for the seven heavy metals are naturally dependent on many variables: the purity of the chemicals used; the purity of the laboratory air; the level of the analytical trace technique employed, etc. Under clean room conditions detection limits of $0.05 \, \mu g \, l^{-1}$ were achieved for all seven elements examined.

All analyses were carried out with the multiple standard addition (here: with threefold standard addition performed automatically) to conform with the usual practice in demanding ultra-trace determinations.

This leads to a total of four curves per heavy metal. Concentrations found were as given in Table 2.22.

Table 2.22.

	$\mu g \, l^{-1}$
Zinc	2.51
Cadmium	<0.05
Lead	1.27
Copper	1.04
Iron	1.31
Nickel	2.64
Cobalt	<0.05

The determination was carried out in three stages:

- Sample in sodium fluoride solution $c(NaF) = 0.01 \, mol/l$
  Zn, Cd, Pb
  pH = 7. Differential pulse anodic stripping voltammetry
  (DPASV) $\Delta u_p = +50 \, mV$
- Cu, Fe
  pH = 7; addition of 1,2-dihydroxybenzene (catechol complex)
  Differential pulse cathodic stripping voltammetry (DPCSW)
  $\Delta u_p = -50 \, mV$
- Ni, Co
  pH = 9, addition of dimethylglyoxime
  (DMG complex); differential pulse cathodic stripping
  voltammetry (DPCSV) $\Delta u_p = -75 \, mV$

Determination of nitriloacetic acid in river water

The technique has been used to determine down to $9.5 \, \mu g \, l^{-1}$ nitriloacetic and in river water samples by stripping voltammetry using the Metrohm MO 646 VA processor and 647 VA stand (Guerrieri and Bucci 1985).

Down to $10 \, \mu g \, l^{-1}$ could be detected and, with slight modifications to the procedure, down to $2 \, \mu g \, l^{-1}$.

Other applications of polarography in the water laboratory include uncomplexed and organically complexed lead, copper and zinc sediments and sludges (Schlosser and Schwedt 1985), copper species in estuarine waters (Nelson 1985), chromium VI in wastewaters and industrial effluents (Horzdorf and Jansen 1984), copper III – organic interactions in estuarine waters (Nelson 1985b), phthalates in wastewaters (Tanaka and Takeshita 1984).

It is noteworthy that in none of these instances could the analysis have been carried out by atomic absorption of inductively coupled plasma techniques, indicating that polarography is finding itself an indispensable place in the armoury of the water chemist.

3 High-purity laboratory water

Laboratory water purification has undergone dramatic changes in the last decade. Chemists, life scientists, and medical technologists are now routinely concerned with impurity levels impossible to measure ten years ago.

As a result, distilled water purity is marginal for much of today's analytical work (Table 3.1). In addition, soaring energy and maintenance costs are making distillation a poor choice. Today, reverse osmosis, deionization, carbon adsorption and membrane microfiltration are all, in some respects, superior to distillation.

To enable users of high-purity water to define their needs more precisely, several professional organizations have established standards for certain classes of use.

With this system it is possible to group laboratory water purification systems according to the classes of use.

Laboratory grade water: types III, IV

Water of this quality has been prepared traditionally by single stage distillation (Table 3.1A). It is adequate for most general laboratory use including qualitative analysis, non-critical media and reagents and glassware washing. Another use for type III/IV water is the pre-treatment of water prior to reagent grade polishing. Reverse osmosis is now a more practical choice than distillation for general laboratory water (Table 3.1B). It is far less expensive, more dependable and almost maintenance-free because contaminants are continuously flushed away with the reject stream.

Analytical grade water: type II

This grade of water has a resistivity of at least 1 MΩ cm at 25°C and is suitable for all but the most critical procedures including spectrophotometry, liquid–liquid extraction potentiometry and volumetric analysis. Traditionally, water of this quality has been produced by single and double distillation. Increasingly, nowadays, this quality is achieved by equipment based on a combination of reverse osmosis and ion-exchange and final microfiltration (Table 3.1C).

Water produced by this method is not of the highest quality and cannot be classified as ultra-high quality. Whereas, for example, it might suffice for use

Table 3.1. *Water purification units*

Supplier	Model	Sensitivity (MΩ cm)	Conductivity (μs cm^{-1})	Applications
A. Distilled water				
Manestry	L4 still	–	–	Pyrogen-free General laboratory use
Hamilton	stills	–	–	General laboratory use
S Bibby	Aquatron	–	0.1–0.2	General laboratory use
Fistreem	Cyclon	–	≤1	Pyrogen-free General laboratory use (can be used with the Cyclon deionizer unit to provide feed water to the still)
Jencons	Autostill range	–	0.5–2.0	General laboratory use
B. Reverse osmosis water				
Fistreem	R060	–	–	General laboratory use and source of water for polishing to high-purity water using the Fistreem Cyclon Unit
Elga	Prima	–	–	General laboratory use and source of water for polishing to high-purity water using the Elga stat UHP or UHQ
Millipore	Milliro system	18 at 25°C	–	General laboratory use and source of water for polishing to high-purity water using the Millipore Milli-Q-system

C. Reverse osmosis – deionization water

Elga	Elgastat Spectrum	Up to 18	0.05	HPLC atomic absorption spectrometry; tissue culture, spectrophotometry, flame photometry. Reference and buffer solutions: general laboratory use
Millipore	Milli R/Q	2–5	–	Spectrophotometry liquid–liquid extraction potentiometers, volumetric analysis

D. Resverse osmosis – multi-column water

Gelman	Water 1 system	18	0.055	Atomic absorption spectrometry; emission spectrometry; HPLC
Fistreem	Nonopure Ultrapure Water system	Up to 18	0.055	Atomic absorption spectrometry; HPLC; spectrophotometers: flame photometry; reference and buffer solutions: tissue culture; enzymology, haematology
Elga	Elgastat UHP and Elgastat UHQ	Precede by reverse osmosis: 5 columns for removal of metals, anions, cations, organics, particulates, pyrogens, bacteria and colloids	18 at 25°C	HPLC, ion chromatography, fluorescence analysis, total organic carbon, microbiology
Millipore	Milli-Q system	Precede by reverse osmosis: 5 columns for removal of metals, anions, cations, organics, particulates, pyrogens, bacteria and colloids	18 at 25°C	Atomic absorption spectrometry, HPLC, total organic carbon, enzymology, tissue culture

Table 3.2. *Ultra high-purity water multi-cartridge systems*

Column	Elgastat UHP	Elgastat UHQ	Fistream Nanopure II	Millipore Milli-Cl system
1	Organics removal	Reverse osmosis (built-in)	Organics removal	Particle/bacteria removal
2	Inorganic ion removal	Organics removal	Colloids removal	Organics removal
3	Ultramicro filtration	Inorganics removal	Inorganics removal	Inorganics removal
4	Photo-oxidation (organics removal)	Microfiltration particle/bacteria removal	Ultra-filtration (particle/bacteria removal)	Final trace organics removal
5	Ultrafiltration (particle/bacteria removal)	Photo-oxidation (organics removal)	–	Ultrafiltration (pyrogen removal)
6	–	–	–	Membrane filtration (micro-organics removal)
	Removal or organics		Activated carbon ultraviolet radiation	
	Removal of inorganics		Ion-exchange resin	
	Removal of colloids, micro-organisms, particles		Ultra-microfiltration, macroreticular resins	
	Removal of particles and bacteria		Ultrafiltration on membrane filters	

in the atomic absorption determination of metals in water samples in the water laboratory at the milligram per litre level, it would be insufficiently pure when carrying out analyses at the microgram per litre level or lower. Similarly, the total organic carbon concentration of type II water would preclude its use in total organic carbon determinations in samples at levels below $100 \, \mu g \, l^{-1}$.

Reagent grade water: type I

Type I water has a resistivity of $18 \, M\Omega \, cm$ at 25°C. This type of water is particularly recommended for procedures that require freedom from trace impurities at the very limits of detection, such as atomic absorption analysis at the microgram or nanogram level, UV and IR spectroscopy, voltammetry, specific ion electrodes, HPLC and total organic carbon measurements in the range $10-100 \, \mu g \, l^{-1}$. Such water quality is a fundamental element in biological work.

Producing water of this quality always requires several stages of treatment because no single process is capable of removing all contaminants. In order to take a heavy load off the columns used in such equipment and to enhance their life and to ensure the highest quality of the final product, it is recommended that the input water is first treated by reverse osmosis to first remove gross impurities. Several such systems are summarized in Table 3.1D and Table 3.2.

4 Anions analysis

4.1 Segmented flow analysis (including cations and organic applications)

Segmented flow analysers, alternatively known as autoanalysers, are extensively used in water laboratories for the routine batch determination (up to 80 samples per hour) of a wide range of determinants including those listed in Table 4.1. This data is applicable to equipment supplied by Skalar BV, Holland.

Segmented flow analysis is based on the principle of pumping a liquid through a system of tubing, dividing it by air bubbles into equal parts or segments and then measuring it. In practice the liquid is a reagent to which the sample to be analysed is added; the resulting colour change is then measured by various methods. Samples are analysed in batches and the system is rinsed between each sample to prevent 'carry-over'. Either a single measurement can be made repeatedly to the sample or it can be divided between a number of analytical modules, so that various measurements are made simultaneously from the same sample.

Table 4.1.

Substance	Range (mg l^{-1})
Aluminium	0–0.5
Hydrolised aluminium	0–1
Ammonia	0–5,0–10 and 0–25 (as N)
Anionic detergents	0–10 (as Manoxol OS)
Arsenic	0–1
Bicarbonate	0–0.5 (as CO_2)
Boron	0–2
Calcium	0–100
Carbon dioxide	0–200
COD	0–300 (as O_2)
Cationic detergents	0–500 $(\mu\text{g ml}^{-1})$
Chloride	0–5, 0–60 and 0–500
Chlorine	0–500

Table 4.1. *(Continued)*

Substance	Range (mg l^{-1})
Chromium VI	0–2
Cyanide	0–0,4 and 0–0.25
DOC	0–5 (as C)
Fluoride	0–2
Hydrazine	0–2 (as N_2H_2)
Hydrogen sulphide	0–1 (as H_2S)
Iron II	0–1
Iodine	0–10
Iron III	0–10
Lead	0–1
Magnesium	0–10
Manganese	0–0.25[1]
Nickel	0–0.1 and 0–1
Nitrate plus nitrite	0–0.5, 0–5, 0–20, 0–40 and 0–100 (as N)
Nitrite	0–0.1 and 0.2[1] (as N)
Non-ionic detergents	0–10 (as Nonyl phenol C_{lo})
pH	4.0–8.0
PV	0–0.1 (as K mnO4)
Phenols	0–2 and 0–200 (as C_6H_5OH)
Hydrolysed phosphate	0–5 (as P)
Ortho total phosphate	0–1, 0–200 (as P)
Orthophosphate	0–0.1 (as P)
Total phosphate	0–25 (as P)
Potassium	0–200
Silicate	0–5, 0–100 (as Si)
Sodium	0–200
Sulphate	0–60, 0–3000 (as SO_4)
Sulphite	0–25 (as SO_3)
Sulphur dioxide	0–25 (as SO_2)
Total amino acids	0–1 and 0–100 (as alanine)
Total alkalinity	0–500 (as C_uCO_3)
Total hardness	0–1°D and 0–50°D
Total iron	0–0.85 and 0.5
Total nitrogen	0–85 (as N)
Urea	0–4 (as CH_4N_2O)
Zinc	0–10

[1] Also in sea water.

The advantages of segmented flow analysis are as follows:

● Many different analytical measurements can be made on a single sample at the same time
● All analytical results are produced in analogue forms if necessary. This reduces sample administration work and transcription errors
● Small sample and reagent volumes are required, effectively reducing costs.

The Skalar analyser unit consists of five components, a sampler, a pump unit, an analytical section, a detection unit and a calculator/computer (Figure 4.1).

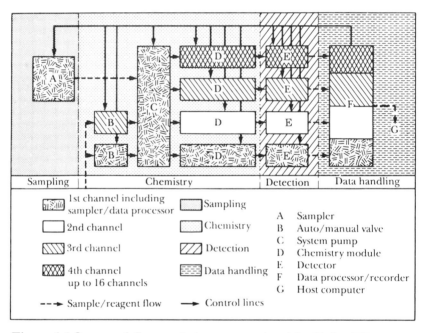

Figure 4.1 *Segmented flow analysis system employed by Skalar BV*

Detections based on spectrophotometry, flame photometry (sodium, potassium) atomic absorption spectrometry (metals) and ion selective electrodes (chloride, fluoride) are available.

Other companies, e.g. Chemlab in UK and Technicon Corporation in USA produce segmented flow analysers. The continuous flow system supplied by Chemlab has been in use in water laboratories for several years.

The original Chemlab system (CAA I) utilizes separate dialysis and high-temperature baths and requires larger volumes of samples and reagents, whereas in the second series system (CAA II) the amounts of samples and reagents required are smaller, due to the heating and dialysis baths being miniaturized and fitted, with all the requisite glassware, into a neat, compact, analytical cartridge. A separate cartridge is constructed for each constituent in the sample to be analysed. Both systems are in current use although the CAA II system is the one most frequently used because it uses smaller volumes of samples and reagents and is more compact.

The ChemLab Flow Analysis Systems give a graphical output on the recorders. If a number of channels are being used the results can be automatically computed and printed out by a data processor. This instrument can be connected to up to eight separate analytical channels. If further statistical analysis of these results is required the data processor can be connected to a mainframe computer by various interfaces.

ChemLab System 4 is a new analytical system for the user who wants to perform single or multiple analyses of a sample with maximum efficiency. The analyser is designed in modular form which allows the user to select the system that best suits his needs and yet can be easily changed or upgraded as the workload alters. System 4 is fast, accurate and precise with a wide range of chemistries available.

4.2 Automatic titration

The titration process has been automated so that batches of samples can be titrated non-manually and the data processed and reported via printouts and screens. Available automatic titrators are reviewed in Appendix 1. One such instrument is the Metrohm 670 titroprocessor. This incorporates a built-in control unit and sample changer so that up to nine samples can be automatically titrated. The 670 titroprocessor offers incremental titrations with variable or constant-volume steps (dynamic or monotonic titration). The measured value transfer in these titrations is either drift controlled (equilibrium titration) or effected after a fixed waiting time; pK determinations and fixed end points (e.g. for specified standard procedures) are naturally included. End-point titrations can also be carried out.

Sixteen freely programmable computational formulae with assignment of the calculation parameters and units, mean-value calculations and arithmetic of one titration to another (via common variables) are available. Results can be calculated without any limitations.

The 670 titroprocessor can also be used to solve complex analytical tasks. In addition to various auxiliary functions which can be freely programmed, up to four different titrations can be performed on a single sample.

In addition to the fully automated 670 system, Metrohm also supply simpler units with more limited facilities which nevertheless are suitable for more simple titrations. Thus the model 682 titroprocessor is recommended for routine titrations with automatic equivalence pointer cognition or to preset end points. The 686 titroprocessor is a lower-cost version of the above instrument again with automatic equivalence point recognition and titration to preset end points.

Mettler produce two automatic titrimeters suitable for use in the water laboratory, the DL 40 GP memotitrator and the lower-cost DL 20 compact titrator. Features available on the LD 40GP include absolute and relative end-point titrations, equivalence point titrations, back-titration techniques, multi-method applications, dual titration, pH stating, automatic learn titrations, automatic determination of standard deviation and means, series titrations, correction to printer, acid balance analogue output for recorder and correction to the laboratory information system. Up to 40 freely definable methods can be handled and up to 20 reagents held on store. Six control principles can be invoked. The DL 20 can carry out absolute (not relative) end-point titrations and equivalence point titrations, back-titration, series titrations, and correction to printer and balance and the laboratory information system. Only one freely definable method is available. Four control principles can be invoked.

The DL 40GP can handle potentiometric, voltammetric or photometric titrations.

Automatic sample changers

Mitsubishi supply an automatic sample changer Model 9T-5GC for combination with their automatic titration model GT-05 to enable automatic titration of multiple samples. Up to 12, 24 and 36 samples can be prepared. Features of this instrument are as follows:

- Various pretreatment (solvent dispensing, sample dissolution, chemical reaction) can be performed by sequential operations
- Reconditioning function for the electrode can be programmed
- Automatic power switch-off function makes unattended operation possible and safe.

4.3 Flow-injection analysis

Flow-injection analysis (FIA) is a rapidly growing analytical technique. Since the introduction of the original concept by Ruzicka and Hansen (1975) in 1975 about 1000 papers have been published, about one third of which deal with the analysis of water.

Flow-injection analysis is based on introduction of a defined volume of sample into a carrier (or reagent) stream. This results in a sample plug bracketed by carrier (Figure 4.2(a)).

The carrier stream is merged with a reagent stream to obtain a chemical reaction between the sample and the reagent. The total stream then flows through a detector (Figure 4.2(b)). Although spectrophotometry is the

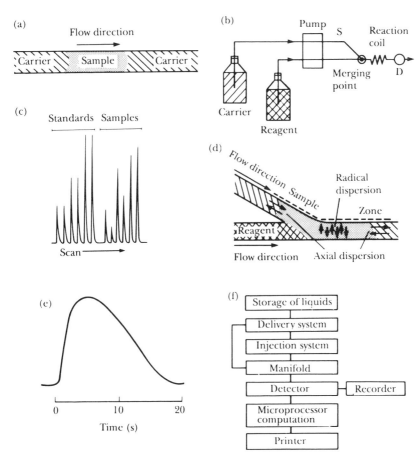

Figure 4.2 *(a) Schematic diagram of the flow pattern in an FIA system directly after injection of sample; (b) simple FIA system for one reagent S denotes the sample injection site and D is the flow through detector; (c) typical FIA peaks (detector output signals); (d) radial and axial dispersion in an injected sample plug; (e) rapid scan of an FIA curve; (f) configuration of an FIA system*

commonly used detector system in this application, other types of detectors have been used, viz fluorimetric, atomic absorption emission spectroscopy and electrochemical (e.g. ion selective electrodes).

The pump provides constant flow and no compressible air segments are present in the system. As a result the residence time of the sample in the system is absolutely constant. As it moves towards the detector the sample is mixed with both carrier and reagent. The degree of dispersion (or dilution) of the sample can be controlled by varying a number of factors, such as sample volume, length and diameter of mixing coils and flow rates.

When the dispersed sample zone reaches the detector, neither the chemical reaction nor the dispersion process has reached a steady state. However, experimental conditions are held identical for both samples and standards in terms of constant residence time, constant temperature and constant dispersion. The sample concentration can thus be evaluated against appropriate standards injected in the same manner as samples (Figure 4.2(c)).

The short distance between the injection site and the merging point ensures negligible dispersion of the sample in this part of the system. This means that sample and reagent are mixed in equal proportions at the merging point.

The mixing technique can be best understood by having a closer look at the hydrodynamic conditions in and around the merging point (Figure 4.2(d)).

In Figure 4.2(d) the hydrodynamic behaviour is simplified in order to explain the mixing process. Let us assume tha there is no axial dispersion and that radial dispersion is complete when the sampler reaches the detector. The volume of the sample zone is thus 200 μl after the merging point (100 μl sample + 100 μl reagent as flow rates are equal). The total flow rate is $2.0 \, ml \, min^{-1}$. Simple mathematics then gives a residence time of 6 s for the sample in the detector flow cell. In reality, response curves reflect some axial dispersion. A rapid scan curve is shown in Figure 4.2(e). The baseline is reached within 20 seconds. This makes it possible to run three samples per minute and obtain baseline readings between each sample (no carry-over) i.e. 180 samples per hour.

The configuration of an FIA system is shown schematically in Figure 4.2(f). The (degassed) carrier and reagent solution(s) must be transported in a pulse-free transport system and at constant rate through narrow Teflon (Du Pont) tubing.

In a practical FIA system peristaltic pumps are usually used since they have several channels and different flow rates may be achieved by selection of a pump tube with a suitable inner diameter.

A manifold provides the means of bringing together the fluid lines and allowing rinsing and chemical reaction to take place in a controlled way. Manifolds with several lines can be assembled as required. These manifolds are mounted on plastic trays and allow the use of different reaction coils.

4.3.1 Instrumentation

Flow-injection analysers available range from relatively low-cost unsophisticated instruments such as those supplied by Advanced Medical Supplies, Skalar and ChemLab to the very sophisticated instruments such as the FIA star 5010 and 5020 supplied by Tecator (Table 4.3).

Table 4.2.

Aluminium	Albumin	Ammonia	Alkalinity	Alkalinity
Ammonia	Alkaline-phosphatase	Glucose	Aluminium	Aluminium
Boron	Bilirubin	Iron	Ammonia	Ammonia
Chloride	Calcium	Nitrate/nitrite	Boron	Calcium hardness
Nitrate/nitrite	Carbon dioxide	Phosphate	Chloride	Chloride
Phosphate	Chloride	Potassium	Chromate	Chromate
Potassium	Cholesterol	Sodium	Copper	Copper
Urea	Creatinine	Reducing sugars	Free cyanide	Free cyanide
	Glucose	Total sugars	Total hardness	Total hardness
	γ-Glutamyl-Transpeptidase		Hydrazine	Hydrazine
	Hydroxyproline		Iron	Iron
	Iron		Nicotine	Nitrate/nitrite
	Magnesium		Nitrate/nitrite	Phosphate
	Phosphate		Phosphate	Silicate
	Potassium		Potassium	
	Protein		Silicate	
	Sodium		Reducing sugars	
	Urea (BUN)		Total sugars	
	Uric acid		Urea	

Table 4.3. *Equipment for flow-injection analysis*

Supplier	Model	Features	Detectors available
Advanced Medical Supplies	LGC 1	Relatively low-cost instrument, recorder output. No computerization on data processing (8 channels)	Colorimeter (other detectors can be used but are not linked in, e.g. atomic absorption, fluorimeter ion-selective electrodes)
Chemlab	–	Relatively low-cost, recorder output or data analysis by microprocessor (3 channels)	Colorimeter
Skalar	–	Relatively low-cost, recorder output on data analysis by microprocessor also carries out segmented flow analysis	Colorimeter, flow cells for fluorimeter and ion-selective electrodes available
Fiatron	Finlite 600	Laboratory process control and pilot plant instrument computerized	Colorimeter
	Fiatrode 400 Fiatrode 410 Fiatrode 430	Flow through analyser/controller, process control analyser	pH and ion-selective electrode
Tecator	FIA star 5025	Relatively low cost manual instrument specifically designed for fluoride, cyanide, potassium, iodide, etc.	Specially designed for use with ion-selective electrodes
	FIA star 5032	relatively low-cost manual instrument (400–700 nm)	Spectrophotometer and/or photometer detectors

Aquatec	Modular, semi or fully automatic operation. Microprocessor controlled. A dedicated instrument designed for water analysis, i.e. dedicated method cassettes for phosphate and chloride, 60–100 samples per hour	Flow through spectrophotometer (400–700 nm)
FIA star 5010	Modular, semi- or fully automatic operation. May be operated with process controller microprocessor. Can be set up in various combinations with 5017 sampler and superflow software which is designed to run on IBM PC/XT computer; 60–180 samples per hour. Dialyses for in-line sample preparation and in-line solvent extraction. Thermostat to speed up reactions	Spectrophotometer (400–700 nm) or photometer can be connected to any flow through detector, e.g. UV/visible, inductively coupled plasma, atomic absorption spectrometer and ion-selective electrodes
FIA star 5020	As above, top of the market, higher sample throughput (up to 300 samples per hour) microprocessor controlled functions, automatic calculation of results, digital presentation of results, automatic recalibration, stopped flow and intermittent pumping for slow reactions, 100-sample sampler, 5 chemifolds, gas diffusion measurements dialysis and solvent extraction. Non-aqueous and corrosive reagents	As FIA star 5010

The following ions can be determined by flow inspection analysis, fluoride, cyanide, potassium, iodide, calcium, ammonium, iron, chromium, phosphate, carbonate, total alkalinity, nitrate, nitrite; also alkalinity, boron, calcium, colour, cyanide, hardness, total nitrogen, phenol, phosporus, FIA star surfactants, Kjeldahl nitrogen, urea (Tecator 5010 analyser).

4.3.2 Applications

Some examples of waget analysis illustrating the potential of this technique are discussed below.

Automated continuous monitoring of inorganic and total mercury in waste water

Bernie (1988) has described an automatic continuous monitoring system for the determination of inorganic and total mercury by flow-injection analysis followed by cold vapour atomic absorption spectrometry. Mercury is removed by aeration from the flowing stream in a specially designed air–liquid separator and swept into a silica cell for absorption measurement at a wavelength of 253.7 μm. The calibration curve extended up to $10 \, \text{mg} \, \text{l}^{-1}$ mercury with a detection limit of $0.02 \, \text{mg} \, \text{l}^{-1}$ mercury.

Heavy metals in sea water

Olsen *et al.* (1983) combined a Tecator flow injection analyser with an atomic absorption spectrometer to determine trace amounts of cadmium, lead, copper and zinc in polluted sea water. They used a simple flow-injection system to inject the samples directly into the flow-injection analysis equipment and were able to analyse samples at the rate of 180–250 samples per hour.

Silica immobilized 8-quinoline was used as a preconcentration tool in flow-injection atomic absorption spectrometry determinations of copper II in some Environmental Protection Agency standard water samples. Very good agreement was obtained with reported values even in very complex sample matrices.

Speciation studies

One of the claimed advantages for flow-injection analysis is a study of speciation of anions and cations in water samples. As an example of this Ruz *et al.* (1986) speciated the different oxidation states of chromium. They were able to obtain a concentration profile for the chromium III and chromium VI species, $HCrO_4^- - Cr_2O_7^{2-}$ and CrO_4^{2-}.

They also developed on-line flow-injection analysis preconcentration methods using a microcolumn of Chelex 100 resin which enabled them to determine lead in sea water at concentrations down to $10 \, \mu\text{g} \, \text{l}^{-1}$ and cadmium and zinc down to $1 \, \mu\text{g} \, \text{l}^{-1}$ at a sampling rate of 30–60 samples per hour. Fang *et al.* (1984) also used a flow-injection system comprising on-line preconcentration on an ion-exchange resin for the determination of heavy metals (copper, zinc, lead and cadmium) in sea water by atomic absorption spectrometry. The respective detection limits were 0.07, 0.03, 0.5 and 0.05 $\mu g\,l^{-1}$.

Chromium in water

Marshall and Mottola (1985) evaluated a method for determining the ionic forms of chromium in water:

Chromium VI
$$H_2CrO_4 \rightleftharpoons HCrO_{4-}$$
$$HCrO_4^- \rightleftharpoons CrO_4^{2-} + H^+$$
$$CR_2O_7^{2-} + H_2O - \rightleftharpoons 2HCrO_4^-$$

Chromium III
$$Cr^{3+} + H_2O \rightleftharpoons Cr(OH)^{2+} + H^+$$
$$Cr(OH)^{2+} + H_2O \rightleftharpoons Cr(OH)_2^+ + H^+$$
$$Cr^{3+} + 3OH^- = \rightleftharpoons Cr(OH)_3$$
$$Cr^{3+} + 4OH^- \rightleftharpoons Cr(OH)_4^-$$

Other recent applications of flow-injection analysis include the computerized simultaneous determination of calcium and magnesium (Canate and Rios 1987b) and of pH, alkalinity and total ionic concentration (Canate and Rios 1987a).

Organic applications

Fiatron supply automated carbohydrate analyser modules Fiazyme 500 series) for the determination of Krebs circle constituents (glucose, fructose, sucrose, lactose, starch, ethanol and 1-lactate).

4.4 Ion-selective electrodes

Inorganic and organic applications

Ion-selective electrode technology is based on the simple measuring principle consisting of a reference electrode and a suitable sensing or indicator electrode sample solution (for the ion being dipped) dipped in the sample solution and connected by a sensitive voltameter. The sensing electrode responds to a difference between the composition of the solution inside and outside the electrode and requires a reference electrode to complete the circuit.

The Nerst equation, $E = E_0 + S \log C$, which gives the relationship between the activity or concentration (C) contains two terms which are constant for a particular electrode. These are E_0 (a term based on the potentials which remain constant for a particular sensing/reference electrode pair) and the slope S (which is a function of the sign and valency of the ion being sensed and the temperature). In direct potentiometry, it has to be assumed that the electrode response follows the Nernst equation in the sample matrix and in the range of measurement. E_0 and slope are determined by measuring the electrode potential in two standard solutions of known composition and the activity of the ion in the unknown sample is then calculated from the electrode potential measured in the sample.

Reference electrodes of interest to the water chemist are of two types – single function and double function. Indicating or sensing electrodes are of four types:

- solid state   Determination of Br^{1-}, Cd^{2+}, Cl^{1-}, Cu^{2+}, CN^{1-}, F^{1-}. I^{1-}. Pb^{2+}, Redox, silver/sulphide $^{1-}$, Na^{+}, CNS^{1-}
- liquid membrane Ca^{2+}, divalent (hardness), fluoroborate, NO_3^{1-}, K^{1+}, ClO_4^{1-}, HF, surfactants
- residual chlorine
- glass sodium

Variables which effect precise measurement by ion selective electrodes are the following:

- concentration range
- ionic strength – an ionic strength adjuster is added to the samples and standards to minimize differences in ionic strength
- temperature
- pH
- stirring
- interferences
- complexation

Traditionally electrodes have been used in two basic ways, direct potentiometry and potentiometric titration. Direct potentiometry is usually used for pH measurement and for measurement of ions like sodium, fluoride, nitrate and ammonium, for which good selective electrodes exist.

Direct potentiometry is usually done by manually preparing ionic activity standards and recording electrode potential in millivolts, using a high-impedance millivoltmeter and plotting a calibration graph on semilogarithmic graph paper (or using a direct reading pH/ion meter which plots the calibration graph internally).

In potentiometric titration techniques, the electrode is simply used to determine the end-point of a titration, much as a coloured indicator would be used.

Direct potentiometry is an accurate technique but the precision and repeatability are limited because there is only one data point. Electrodes drift and potential can rarely be reproduced to better than ± 0.5 mV so that the best possible repeatability in direct measurement is usually considered to be $\pm 2\%$.

Orion, the leading manufacturers of ion-selective electrodes, supply equipment for both direct potentiometry (EA 940, EA 920, SA 720 and SA 270 meters) and potentiometric titration (Orion 90 autochemistry system).

Table 4.4. *Orion pH/ISE meter features chart*

Feature	Orion pH/ISE meters			
	EA 940	EA 920	SA 720	SA 270
pH	√	√	√	
Direct concentration readout in any unit	√	√	√	√
mV	√	√	√	√
Rel mV	√	√	√	
Temperature	√	√	√	√
Oxygen	√	√		√
Redox	√	√	√	√
Dual electrode inputs	√	√		
Expandable/upgradable	√	√		
Automatic anion/cation electrode recognition	√	√	√	√
Multiple point calibration	√			
Incremental analytical techniques	√	√[1]		
Multiple electrode memory	√	√		
Prompting	√	√	√	√
Ready indicator	√	√	√	
Resolution and significant digit selection	√	√	√	√
pH autocal	√	√	√	
Blank correction	√	√		
Multiple print option	√	√		
Recorder output	√	√	√	
RS-232C output	√	√	√	
Adjustable ISO	√	√	√	√
Automatic temperature compensation-line and battery operation	√	√	√	√

[1] With PROM upgrade.

4.4.1 Instrumentation

Ion-selective electrode equipment: Orion direct potentiometry meters

A review of these four meters, in Table 4.4, shows that only the EA 940 has a facility for multiple point calibration and this places it at the top of their range of direct-potentiometry meters. This instrument automatically prints out results. It has a memory for storing calibration information for all the electrodes.

The EA 920 is a lower-priced instrument for two-step calibration. It also has a memory for storing calibration information. The SA 720 and the portable SA 270 are relatively inexpensive bottom of range instruments with more limited capabilities.

The Orion 960 autochemistry system (direct potentiometry –
potentiometric titration)
This is the top-of-the-range instrument. In addition to direct potentiometry and potentiometric titrations it has other features not previously incorporated in potentiometric analysers.

The 960 uses twelve basic analytical techniques. To do an analysis one of these techniques is chosen and modified to suit the requirements of the particular sample. The memory will accommodate up to 20 methods.

KAP^{TM} analysis is a time-saving technique that eliminates sample preparation and calibration. Simply weigh sample into a beaker, add water and measure. Aliquots of one standardizing solution or reagent are added automatically to the simple and sample concentration is determined from the changes in potential observed after each addition. Every step is performed in one beaker.

Results from KAP analysis are automatically verified in two ways. First a check for electrode drift and noise is performed at the beginning of each analysis. Second, each sample is spiked with standard as part of the analysis and recovery of the spike is calculated.

GAP^{TM} analysis is a faster way to perform many titrations. GAP analysis actually predicts the location of the end point so there is no need to titrate all the way. And GAP analysis allows titrations to be preformed at low levels where conventional techniques yield very poor end-point breaks or asymmetrical curves.

HELP analysis is a diagnostic technique in which the instrument studies the data collected and recommends the optimum procedure for repetitive analysis of similar samples.

The heart of the Orion 960 autochemistry system is the EA 940 expandable analyser – an advanced pH/ISE meter.

Orion supply both electrodes and measuring equipment. Ingold, on the other hand, supply only electrodes. EDT Analytical (UK) also manufacture ion-selective electrodes.

4.4.2 Applications

Some water analysis methods published by Orion are reviewed in Table 4.5.

4.5 Ion chromatography

Anion, cation and organic applications

Ion chromatography was originally developed by Small *et al.* (1975) for rapid and sensitive analysis of inorganic anions, such as sulphate, chloride and nitrate using specialized ion-exchange columns and chemically suppressed conductivity detection. Advances in column and detection technologies have expanded this capability to include a wider range of anions as well as organic ions and metals. These recent developments, discussed below, provide the water chemist with a means of solving many problems that are difficult, if not impossible, using other instrumental methods. Ion chromatography can analyse a wide variety of non-metals more easily than either atomic absorption spectrometry or inductively coupled plasma techniques. These include halides, ammonia, nitrate, phosphate, sulphate, borate and silicate.

Metals determination is an excellent example of the problem-solving power of ion chromatography. As a stand-alone instrument, an ion chromatograph offers several advantages over atomic absorption spectrometry and inductively coupled plasma spectroscopy. These include oxidation state speciation, metal sampler analysis, sea water analysis and the determination of trace metals in strong acids for analysis of acid digests of sediments or biota. Ion chromatography can complement atomic absorption and plasma techniques as a back-up technique and as an alternative to wet chemistry for cross checking results or as a coupled method where the interference-free metal band from the ion chromatograph is coupled to an atomic absorption spectrometer or an inductively coupled plasma system to improve sensitivity and selectivity.

At the heart of the ion chromatography system is an analytical column containing an ion-exchange column on which various anions and/or cations are separated before being detected and quantified by various detection techniques such as spectrophotometry, atomic absorption spectrometry (metals) or conductivity (anions).

Ion chromatography is not restricted to the separate analysis of only anions or cations, with the proper selection of the eluent and separator columns the technique can be used for the simultaneous analysis of both anions and cations.

Table 4.5. *Applications of orion ion-selective electrodes*

Determination	RSD (%)	Orion instrument	Technique
Inorganic applications			
Sulphate in river water	0.9	Orion 960	First derivative titration
Free cyanide in waste water	2.5–5.4	Orion 960	KAP analysis (titration or first derivative titration)
Water hardness	0.8–1.0	Orion 960	First derivative titration
Residual chlorine in tap water	2.1	Orion 960	KAP analysis (titration)
Chloride			KAP analysis (titration)
Alkalinity	–	Orion 960	First-derivative titration
Organic applications			
Nitriloacetic acid in process waters	–	Orion 960	Titration
Surfactants in water	–	Orion 960	Titration
Phenol and sodium hydroxide in process water	–	Orion 960	Titration

Separations of anions

This original method for the analysis of mixtures of anions used two columns attached in series packed with ion-exchange resins to separate the ions of interest and suppress the conductance of the eluent, leaving only the species of interest as the major conducting species in the solution. Once the ions were separated and the eluent suppressed, the solution entered a conductivity cell, where the species of interest were detected.

The analytical column is used in conjunction with two other columns, a guard column which protects the analytical column from troublesome contaminants, and a pre-concentration column.

The intended function of the pre-concentration column is twofold. First, it concentrates the metallic ions present in the sample, enabling very low levels

Table 4.5. (*Continued*)

Potentiometric applications (Orion EA 964 and EA 920)

In natural water	In boiler feedwater	In sea water	In waste water	In reactor coolants	In potable water	In rain water
NH_4^{1+}	NH_4^{3+}	NH_4^{1+}	NH_4	As^{3+}	F^{1-}	F^{1-}
Ca^{2+}	Ca^{2+}	Cl^{1-}	Cl^{1-}	BOD	NO_3^{1-}	
Cl^{1-}	Cl^{1-}	F^{1-}	CN^{1-}	B^{3+}	Cl^{1-}	
Cu^{2+}		NO_3^{1-}	NO_3^{1-}			
CNO^{1-}			Resid Cl			
CN^{1-}			dis oxygen			
CN^{1+}						
F^{1-}						
M^{1+}						
Hg^{1+}						
Ni^{2+}						
NTA						
NO_2^{1-}						
Protein						
S^{2-}						
S						
Resid Cl						
dis oxygen						

Other ions that can be determined by ion selective electrode but water analysis not specifically mentioned:

Cd^{2+}	Ag^{1+}/S^{2-}	Br^{1-}
I^{1-}	Na^{1+}	Ag^{1+}
Pb^{2+}	SCN^{1-}	Ba^{2+}
ClO_4^{1-}	Hardness	SO_2
K^{1+}		CO_2 and nitrogen oxides

of contaminants to be detected. Second, it retains non-complexed ions on the resin, while allowing complexed species to pass through.

Separation of metals (Rubin and Heberling 1987)

A liquid sample is introduced at the top of the ion exchange analytical column (the separator column, Figure 4.3). An eluent (containing a complexing agent

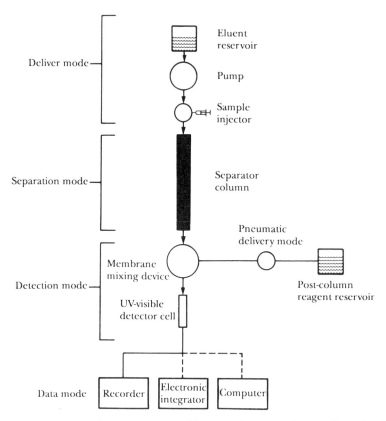

Figure 4.3 *Ion chromatography with post-column reaction configuration for metals analysis*

in the case of metal determination) is pumped through the system. This causes the ionic species (metal ions) to move through the column at rates determined by their affinity for the column resin. The differential migration of the ions allows them to separate into discrete bands.

As these bands move through the column, they are delivered, one at a time, into the detection system. For metals, this comprises a post-column reactor that combines a colouring reagent (i.e. pyridyl azoresorcinol, PAR) with the metal bands. The coloured bands can then be detected by the appropriate detection mode. In the case of metal PAR complex detection, visible wavelength absorbance is employed.

The detector is set to measure the complexed metal band at a preselected wavelength. The results appear in the form of a chromatogram, essentially a plot of time the band was retained on the column versus the signal it produces in the detector (Figure 4.4(a)). Each metal in the sample can be identified and

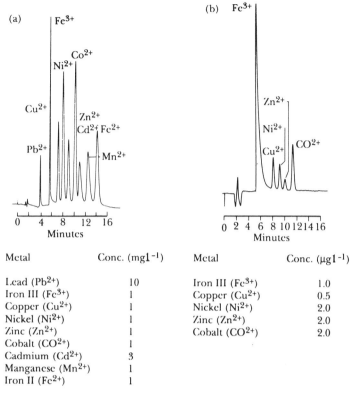

Figure 4.4 *Ion chromatography: (a) determination of nine transition metal ions; (b) trace metals in 0.1 M sulphuric acid (8 ml pre-concentrated)*

Metal	Conc. (mg l^{-1})
Lead (Pb^{2+})	10
Iron III (Fe^{3+})	1
Copper (Cu^{2+})	1
Nickel (Ni^{2+})	1
Zinc (Zn^{2+})	1
Cobalt (CO^{2+})	1
Cadmium (Cd^{2+})	3
Manganese (Mn^{2+})	1
Iron II (Fe^{2+})	1

Metal	Conc. (µg l^{-1})
Iron III (Fe^{3+})	1.0
Copper (Cu^{2+})	0.5
Nickel (Ni^{2+})	2.0
Zinc (Zn^{2+})	2.0
Cobalt (CO^{2+})	2.0

quantified by comparing the chromatogram against that of a standard solution.

Because only the metal ions of interest are detected, ion chromatography is less subject to interferences compared with other methods. And since individual metals and metal compounds form distinct ions with differing retention times, it is possible to analyse several of them in a single run – typically less than 20 minutes.

By selecting the appropriate column for separating the ions of interest in a sample, chemists can now separate and analyse the oxidation state of many metals, determine group I and II metals, metal complexes, and a complete range of inorganic and organic ions in a sample with excellent speed and sensitivity.

Table 4.6 shows a comparison of detection limits of ion chromatography versus flame atomic absorption spectrometry for ideal, single components in de-ionized water. On small-volume injections (50 µl injection) ion chromatography compares well with atomic absorption spectrometry. With sample

Table 4.6. *Metal detection limits by ion chromatography*

Metal species detected by ion chromatography		Detection limit ($\mu g\ l^{-1}$)		
		Direct	*Precon-centrated*	*Flame AA*
Aluminium	Al^{3+}	56	0.5	20
Barium	Ba^{2+}	100	0.1	20
Cadmium	Cd^{2+}	10	0.1	1
Calcium	Ca^{2+}	50	0.5	3
Caesium	Cs^+	100	0.1	20
Chromium	Cr(III) as CrEDTA	1000	10	3
Chromium	Cr(VI) as CrO_4	50	1	3
Cobalt	Co^{2+}	3	0.03	5
Copper	Cu^{2+}	5	0.05	2
Dysprosium	Dy^{3+}	100	1	60
Erbium	Er^{3+}	100	1	60
Europium	Eu^{3+}	100	1	–
Gadolinium	Gd^{3+}	100	1	2000
Gold	Au(I) as $Au(CN)^-$	100	10	10
Gold	Au(III) as $Au(CN)_4^-$	100	10	10
Holmium	Ho^{3+}	100	1	60
Iron	Fe(II)	10	0.1	5
Iron	Fe(III)	3	0.03	5
Lead	Pb^{2+}	100	1	1
Lithium	Li^+	50	0.5	2
Lutetium	Lu^{3+}	100	1	–
Magnesium	Mg^{2+}	50	0.5	0.2
Molybdenum	as MoO_4	50	1	10
Nickel	Ni^{2+}	25	0.3	8
Palladium	as $PdCl_4^{2-}$	10	1	20
Platinum	as $PtCl_6^{2-}$	10	1	50
Potassium	K^+	50	0.5	1
Rubidium	Rb^+	100	1	2
Samarium	Sm^{3+}	100	1	700
Silver	as $Ag(CN)_2^-$	100	10	2
Sodium	Na^+	50	0.5	0.4
Strontium	Sr^{2+}	100	1	6
Terbium	Tb^{3+}	100	1	2000
Thulium	Tm^{3+}	100	1	–
Tin	Sn(II)	100	1	80
Tin	Sn(IV)	100	1	80
Tungsten	as WO_4^{2-}	50	1	100
Uranium	as UO_2^{2+}	5	0.05	7000
Ytterbium	Yb^{3+}	100	1	–
Zinc	Zn^{2+}	10	0.1	0.6

pre-concentration technique (5 ml concentration) the detection limits for ion chromatography can surpass those of graphite furnace atomic absorption spectrometry. In ion chromatography sample pre-concentration can be performed within the system and does not require additional sample handling.

Oxidation state speciation
Ion chromatography can distinguish iron II from iron III, chromium III from chromium IV and gold I from gold III, to name but some examples.

Metal complex analysis
When metals from stable complexes such as cyano, $(Au/CN)_2$, $Au(CN)_4$, $Ag(CN)_2$, $Co(CN)_6$, or EDTA complexes ($MEDTA^{2-}$, where M is Pb, Cu, Zn or Nl), they become distinct ions that can be specifically determined by ion chromatography. For example $Co(CN)_6^{3-}$, $Au(CN)_2$ and $Au(CN)_4^{-}$ can be determined directly. Chromate, molybdate and tungstate are included in this category.

Analysis of sea water
The analysis of metals in the presence of high concentrations of sodium is a common analytical problem. Broad-band spectral interferences from the sodium ion make most trace metal work difficult or impossible with atomic absorption spectrometry. On the other hand, ion chromatography suffers no interference from sodium ions up to about 5% salt solution, i.e. sea water.

Transition metals in strong acids
While high concentration of acids or bases can limit the applicability of atomic absorption spectrometry ion chromatography allows direct injection of up to 10% concentrated acids or bases. This is extremely convenient in the direct analysis of acid digested samples such as digests of river sediments, sea creatures or biological material (Figure 4.4(b)).

Ultra-trace analysis of metals in high purity and rain water
Utilizing ion exchange pre-concentration methods within the chromato-graph, extremely low concentrations of metals in ultra-pure de-ionized water can be measured with ion chromatography. These detection limits are typically in the sub-picogram range.

4.5.1 Instrumentation

Numerous manufacturers now supply instrumentation for ion chromato-graphy (see list of suppliers in Appendix 1). However, Dionex are still the leaders in the field; they have been responsible for many of the innovations introduced into this technique and are continuing such developments.

Dionex series 4000i ion chromatographs

Some of the features of this instrument are tabulated below:

- Chromatography module
- Up to six automated valves made of chemically inert, metal-free material eliminate corrosion and metal contamination
- Liquid flow path is completely compatible with all HPLC solvents
- Electronic valve switching, multidimensional, coupled chromatography or multi-mode operation
- Automated sample clean up or pre-concentration
- Environmentally isolates up to four separator columns and two suppressors for optimal results
- Manual or remote control with Dionex Autoion 300 or Autoion 100 automation accessories
- Individual column temperature control from ambient to 100°C (optional)

Dionex Ion-Pac columns

- Polymer ion exchange columns are packed with new pellicular resins for anion or cation exchange applications
- New 4μ polymer ion exchange columns have maximum efficiency and minimum operating pressure for high-performance ion and liquid chromatography applications
- New ion exclusion columns with bifunctional cation exchange sites offer more selectivity for organic acid separations
- Neutral polymer resins have high surface area for reversed phase ion-pair and ion-suppresion applications without pH restriction
- 5 and 10μ silica columns are optimized for ion-pair, ion suppression and reversed phase applications

Micromembrane suppressor

The micromembrane suppressor makes possible detection of non-UV-absorbing compounds such as inorganic anions and cations, surfactants, antibiotics, fatty acids and amines in ion-exchange and ion-pair chromatography.

Two variants of this exist: the anionic (AMMS) and the cationic (CMMS) suppressor. The micromembrane suppressor consists of a low dead volume eluent flow path through alternating layers of high-capacity ion-exchange screens and ultra-thin ion exchange membranes. Ion-exchange sites in each screen provide a site-to-site pathway for eluent ions to transfer to the membrane for maximum chemical suppression.

Dionex anion and cation micromembrane suppressors transform eluent ions into less conducting species without affecting sample ions under analysis. This improves conductivity detection, sensitivity, specificity, and baseline stability. It also dramatically increases the dynamic range of the system for inorganic and organic ion chromatography. The high ion-exchange capacity of the MMS permits changes in eluent composition by orders of magnitude making gradient ion chromatography possible.

In addition, because of the increased detection specificity provided by the MMS sample, preparation is dramatically reduced, making it possible to analyse most samples after simple filtering and dilution.

Conductivity detector

* High-sensitivity detection of inorganic anions, amines, surfactants, organic acids, Group I and II metals, oxy-metal ions and metal cyanide complexes (used in combination with MMS)
* Bipolar-pulsed excitation eliminates the non-linear response with concentration found in analogue detectors
* Microcomputer-controlled temperature compensation minimizes the baseline drift with changes in room temperature.

UV/Vis detector

* High-sensitivity detection of amino acids, metals, silica, chelating agents, and other UV absorbing compounds using either post-column reagent addition or direct detection
* Non-metallic cell design eliminates corrosion problems
* Filter-based detection with selectable filters from 214 to 800 nm
* Proprietary dual wavelength detection for ninhydrin-detectable amino acids and PAR-detectable transition metals.

Optional detectors

In addition to the detectors shown, Dionex also offers visible, fluorescence, and pulsed amperometric detectors for use with the series 4000i.

Dionex also supply a wide range of alternative instruments, e.g. single channel (2010i) and dual channel (2020i). The latter can be upgraded to an automated system by adding Autoion 100 or Autoion 300 controller to control two independent ion chromatograph systems. They also supply 2000i series equipped with conductivity pulsed amperometric, UV-visible, visible and fluorescence detectors.

4.5.2 Applications

Some typical separations of anions, cations and organics achieved with this system are shown in Figure 4.6. These illustrate the extreme value of this technique to the water chemist.

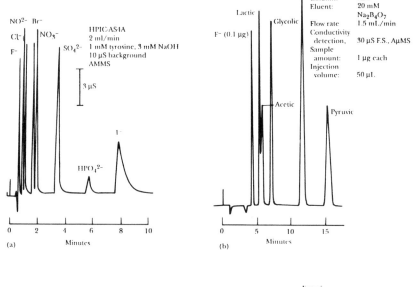

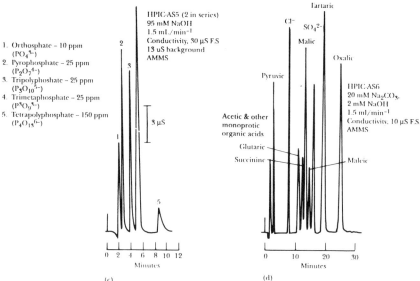

Figure 4.5 *Ion chromatograms obtained with Dionex instrument using (anodic) AMMS and CMMS micromembrane suppression: (a) anions with micromembrane suppressor; (b) monoprotic organic acids by anion exchange; (c) separation of polyphosphates using micromembrane suppressor; (d) diprotic organic acids by anion exchange; (e) simultaneous mono- and divalent cations, isocratic HPIC-CS3, CMMS; (f) monovalent cations HPIC-CS3, CMMS; (g) simultaneous mono- and divalent cations, step gradient HPIC-CS3, CMMS*

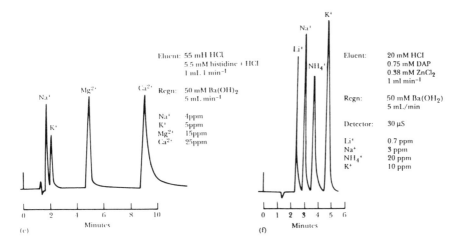

(c)

(f)

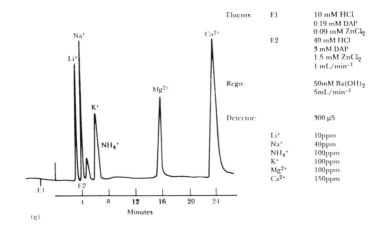

(g)

A further development is the Dionex HPIC AS5A-5μ analytical anion separator column. This offers separation efficiency previously unattainable in ion chromatography. When combined with a gradient pump and an anion micromembrane suppressor the AS5A-5μ provides an impressive profile of inorganic ions and organic acid anions from a single injection of sample (Figure 4.6(a)). Note that phosphate and citrate, strongly retained trivalent ions are efficiently eluted in the same run that also gives baseline resolution of the weakly retained monovalents fluoride, acetate, formate and pyruvate. Quantitation of all the analytes shown in Figure 4.6(b) using conventional columns would require at least three injections under different eluent conditions.

Another benefit of using the AS5A-5μ gradient pump combination is the ability to easily change the order of elution of ions with different valencies

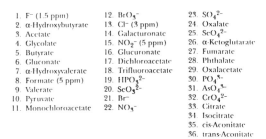

1. F⁻ (1.5 ppm)
2. α-Hydroxybutyrate
3. Acetate
4. Glycolate
5. Butyrate
6. Gluconate
7. α-Hydroxyvalerate
8. Formate (5 ppm)
9. Valerate
10. Pyruvate
11. Monochloroacetate

12. BrO₃⁻
13. Cl⁻ (3 ppm)
14. Galacturonate
15. NO₂⁻ (5 ppm)
16. Glucuronate
17. Dichloroacetate
18. Trifluoroacetate
19. HPO₃²⁻
20. SeO₃²⁻
21. Br⁻
22. NO₃⁻

23. SO₄²⁻
24. Oxalate
25. SeO₄²⁻
26. α-Ketoglutarate
27. Fumarate
28. Phthalate
29. Oxalacetate
30. PO₄³⁻
31. AsO₄³⁻
32. CrO₄²⁻
33. Citrate
34. Isocitrate
35. cis-Aconitate
36. trans-Aconitate

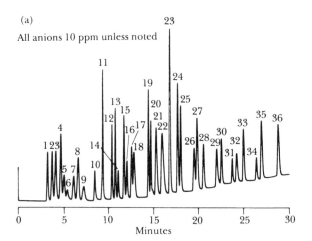

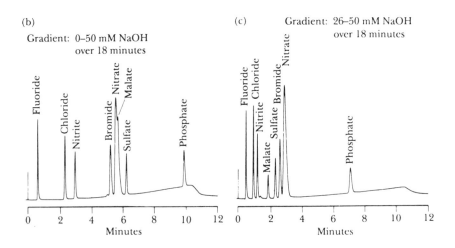

Figure 4.6 *Multi-component analysis by ion chromatography (Dionex)*

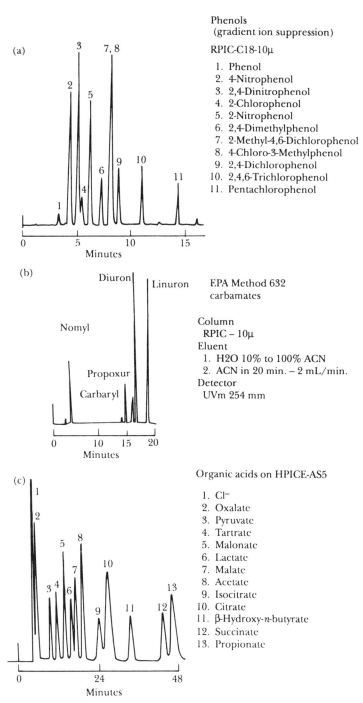

Phenols
(gradient ion suppression)

RPIC-C18-10µ

1. Phenol
2. 4-Nitrophenol
3. 2,4-Dinitrophenol
4. 2-Chlorophenol
5. 2-Nitrophenol
6. 2,4-Dimethylphenol
7. 2-Methyl-4,6-Dichlorophenol
8. 4-Chloro-3-Methylphenol
9. 2,4-Dichlorophenol
10. 2,4,6-Trichlorophenol
11. Pentachlorophenol

EPA Method 632
carbamates

Column
  RPIC – 10µ
Eluent
  1. H2O 10% to 100% ACN
  2. ACN in 20 min. – 2 mL/min.
Detector
  UVm 254 mm

Organic acids on HPICE-AS5

1. Cl⁻
2. Oxalate
3. Pyruvate
4. Tartrate
5. Malonate
6. Lactate
7. Malate
8. Acetate
9. Isocitrate
10. Citrate
11. β-Hydroxy-*n*-butyrate
12. Succinate
13. Propionate

Figure 4.7 *Ion chromatography of (a) phenols; (b) carbamates; (c) organic acids*

simply by changing the gradient profile. For example, if nitrate were present in high enough concentration to interfere with a malate peak, the malate peak could be moved ahead of the nitrate peak by using a slightly different gradient (Figure 4.6(b,c)).

The AS5A-5µ is recommended for use in separations of inorganic ions and organic ions of low molecular weight. It requires a high-pressure pump (Dionex gradient pump) and injection valve (Dionex BF-2 valve) both of which are supplied in Dionex series 4000 i ion chromatographs. An anion micromembrane suppressor (AMMS) is also required and an anion trap column is needed for gradient elution.

A further new column, the HPIC-CS3, is the first analytical cation-exchange separator column that can be used for simultaneous quantitation of group IA and IIA metal ions (Figure 4.5(e,f,g)).

The Dionex AS4A is an analytical anion-exchange separator column designed for optimum performance in separations of common inorganic ions. Seven common anions (F, Cl, NO_2, Br, NO_3, PO_4^{3-} and SO_4^{2-}) can be determined in 6 minutes (Figure 4.5(a)).

Other applications of ion chromatography to water analysis include the determination of anions (Tretter *et al.* 1985) and cations in environmental samples, trace anions and cations in high-purity water (Markos Varga *et al.* 1954) and in rainwater (Rowland 1986), carbonate and bicarbonate in natural water samples (Tanaka and Fritz 1987), chromium III and chromate in waste and tap waters, cyanide in trade effluents and river waters, polynuclear aromatic hydrocarbons, carbamate pesticides (Figure 4.7(b)) phenols (Figure 4.7(a)) and organic acids (Figure 4.7(c)) in natural waters. Ion chromatography has now been adopted as the standard procedure for the determination of chloride, nitrate and sulphate in potable water by the Food and Drug Administration (US Environmental Protection Agency 1984).

5 *Miscellaneous laboratory measurements*

5.1 pH

Portable instruments

It is often necessary to make pH measurements on site, e.g. in field or plant situations, as the pH of samples can sometimes change during the delay encountered between sampling and measurement in the laboratory. This being the case, it is of course, important to look for the same level of accuracy and reliability as one would when choosing a laboratory meter.

Group 1: Basic pH meters

The vast majority of pH measurement applications require the most modest of instruments. This level of meter generally provides manual temperature compensation and offers few facilities other than pH measurement. If the sample and buffer temperature remain constant and the highest level of accuracy is not required, then this type of meter is often most appropriate, especially when the budget is low.

Group 2: General-purpose pH meters

When accuracy of ± 0.01 pH units becomes important then temperature effects do too. The slope of a pH electrode is temperature dependent and if this is not taken into account then the expected error is 0.003 pH unit/K change/pH unit away from calibration.

Temperature compensation involves measuring the temperature of all buffers and samples and dialling or keying in the appropriate values using the facility provided on most general-purpose pH meters. This method is most often used when sample size prevents the use of an extra probe. By far the easiest method of temperature compensation is automatic electronic compensation using a separate temperature probe. This probe senses the solution temperature and automatically feeds the information back to the meter, which compensates for any changes.

Group 3: Quality-control instruments

Accuracy excluded, the most fundamental requirement is to ensure that readings are repeatable under all circumstances and with all staff, regardless of experience.

The microcomputer can be of great value by not only providing greater accuracy but ensuring that a set routine is strictly followed. Without doubt, the microprocessor provides the fastest way to ensure accurate calibration for reliable pH measurements. A good instrument will automatically recognize standard buffers over a wide temperature range and prompt the user to follow a pre-determined routine.

Only a microprocessor-based instrument can stop once and for all errors due to misinterpretation of the reading shown on the display. Sophisticated software can monitor the input signal and, at the appropriate moment, either 'freeze' the display or signal a stable recording. Repeatable readings are thus automated, avoiding operator errors.

In a busy water laboratory with many samples to record, a data output signal can prove to be an invaluable asset, allowing either direct printing of results or storage via a suitable computer. Ensure that the signal is bidirectional for maximum flexibility and not simply a printer output.

Group 4: Research-quality instruments

Versatility and high accuracies are key words here. Once again, microprocessor-based instruments can provide powerful features to give total confidence in the results obtained.

A research quality pH, electrical resistance instrument must provide 0.1 mV resolution for both relative and absolute measurements, to perform complicated analytical techniques such as known addition/subtraction and basic ISE work.

Group 5: Ion-concentration meters

Once again, a microprocessor-based instrument can make ISE work as easy as pH measurements, totally avoiding the need to plot graphs, etc.

The particular requirement for each of the five groups of pH meters listed above are summarized in Table 5.1. There are a vast number of suppliers of pH meters, some of which are listed in the Appendix. Rather than attempt to discuss all of these, some examples are selected below.

In Table 5.1 are listed the models supplied by two leading producers which meet some of the requirements of a pH meter. It is seen that the instruments range from basic (BDT; Beta 51, SI, Beta 52, Beta 53, Orion 611, Jenning 3050, 3070, PH M9) through varying degrees of sophistication (BDT; Beta 500, Micro 2, Orion, SA 250, SA 720, Jenway 3100, PHMMII).

Table 5.1. *Specifications of pH meters produced by EDT and Orion*

Characteristic	EDT	Orion	Fenway portable types
0.01 pH resolution	Beta 51, Beta 52, Beta 53, Beta 500	611. SA 520 SA 250	3100, 307L, 3050, PHM 3, PHM 4, PHM 5, PHM 11
Manual temperature compensation	Beta 51, Beta 52, Beta 53, Beta 500, Sigma 3	–	3050, PHM 3, PHM 5
Autotemperature compensation	Beta 53, Beta 500	SA 720, SA 520, SA 250, SA 230	3100, 3070, 3060, PHM 4, PHM 9, PHM 11
Temperature measurement	Micro 2	SA 720, 611, SA 520, SA 250, SA 230	3100, 3070, PHM 4, PHM 4, PHM 11
mV measurement mV resolution	Beta 51, Beta 52, Beta 53, Beta 500	–	3100, 3070
mV measurement 0.1 mV resolution	Beta 500	–	–
Recorder output	Micro 2	SA 720	–
Printer output	Micro 2	SA 720	–
Autobuffering	Beta 500, Micro 2	SA 720, SA 520, SA 250	3100
Autocalibration	Beta 500, Micro 2	SA 720, SA 520, SA 250	3100
Direct concentration reading	Micro 2	SA 720	3100

A further example of the range of pH meters available is provided by the models 3010, 3020 and 3030 available from Jenway. Model 3010 is a low-cost high-performance instrument capable of simple operation control which reads pH, millivolts and temperature. Model 3020 is microprocessor controlled, having a dual display showing either millivolts or pH. In the pH mode it has automatic calibration and will recognize standard pH buffers and relate calibration to their individual temperature characteristics. The instrument has manual and automatic temperature compensation. The digital display indicates both mode and entry errors. Model 3030 is the top-of-the-range instrument with a powerful minicomputer and high resolution. It has automatic calibration and dual measurement display, and in addition has a two-point alarm/control function. This allows the user to set two alarm set points so that when an alarm threshold is reached an internal audible alarm is sounded and an electrical output is provided for switching functions.

Palintest also produce a low-cost stick pH meter PT105 and a higher-cost minicomputer-controlled pH meter (PT 110) which features automatic temperature compensation, automatic buffer recognition and automatic calibration against buffer solutions. Buffer information is stored in a memory even when the instrument is switched off.

5.2 Temperature

Many of the pH and electrical meters available also measure the temperature of the sample. However, there are also simple measuring devices that record sample temperature only. These are summarized in the Appendix.

Jenway manufacture a full range of hand-held battery-powered probe thermometers which, between them, cover the range -100 to $200°C$.

5.3 Electrical conductivity and resistivity

5.3.1 Conductivity meters

Conductivity meters can generally be divided into either general-purpose types or more sophisticated meters with research capabilities.

Group 1: General-purpose conductivity meters

The perfect instrument would offer accurate measurement of solutions from pure water to sea water with a single cell. Unfortunately no one single cell can cover the wide range of conductivities met in practice.

Important points to look for, therefore, are the availability of various types of conductivity cells and easy calibration procedures (Table 5.2).

Table 5.2.

Conductivity measurement	*Range* (μs)
Ultra-pure water	0–2
Boiler-feed waters	0–20
Portable (drinking water)	0–200
Rinse water	0–2000
Circulated coolant	0–20 000
Sea water	0–2000 000

Cell constants

To improve accuracy at the extreme ends of the conductivity measurement ranges it is desirable to use a conductivity cell with a cell constant other than $1\,cm^{-1}$ (IK″) type that is virtually always supplied as standard with most instruments.

If measuring ultra-pure or distilled water, select a cell with a low constant $(0.01–0.1\,cm^{-1})$, and if measuring high-concentrated solutions, choose a high cell constant of at least $10\,cm^{-1}$. The effect of these cells is to increase the resolution of the meter's display and to provide greater stability.

Temperature compensation

Most general-purpose conductivity meters provide temperature compensation either manually or automatically referred to 20 or 25°C at a fixed coefficient of about $2\%\,K^{-1}$. While this is virtually an international standard, suitable for simple electrolytes, accurate measurements of complex electrolytes require a much more sophisticated meter.

Group 2: Research-quality instruments

Due to the complex nature of analytical conductivity measurement, the latest generation of microprocessor-based instruments should be considered as they can provide facilities that would be time-consuming or impossible to achieve with a conventional instrument.

Resistivity

Ultra-pure water is usually expressed as a megohm value or the reciprocal of conductivity.

Table 5.3. *Characteristics of commercially available electrical conductivity meters*

Characteristics	Research instruments		General-purpose instruments			
	Supplier PDP	Fenway	PDT	Orion	Fenway	Palintest
EC ranges 0–20 µs	Beta 800	4020	Alpha 800	–	3410 3420 4010	BT 115
0–200 µs	Beta 50	PCM 1 4020	Alpha 800[1]	–	3410 4010 3420 4070	PT 115
0–2000 µs	Beta 800	PCM 1 4020	Alpha 800[1]	–	3410 4010 3420 4070	PT 115
0–20 000 µs	Beta 800	4020	Alpha 800[1]	–	3410 4010 3420 4070	–
0–200 000 µs	Beta 800	PCM 1 4020	Alpha 800[1]	–	3410 4010 3420	–

Autoranges	Beta 800	4020	—	—	—	—
Temperature measurement	Beta 800	PCM 1 4020	—	—	4010 4070	—
Manual temperature compensation	Beta 800	4020	Alpha 800	—	4010 3410 3420	PT 115
Autotemp. Compensation	—	PCM 1 4020	Alpha 800 Sigma 3	SL1	4010 4020 4060	—
Cell-constant adjustment	—	PCM 1 4020	Alpha 800	—	3410 4010 3420 4070	—
Alternative cells available	—	PCM 1 4020	Alpha 800	—	3410 4010 3420 4070	—
Autoadjustment of temp. coeff.	Beta 800	4020	—	—	—	—
Printer recorder	Beta 800	4020	—	—	—	—
Microprocessor	Beta 800	4020	—	—	—	—

¹ Most commonly used ranges only.

Total dissolved solids

In a sense, all conductivity meters are TDS meters in that it is the ions such as chlorides, nitrates and sulphates, together with cations such as sodium and calcium, that carry the electrical current making the measurement possible. However TDS as displayed by most instruments is a very arbitrary measurement used mainly in the water-treatment industry. This reading is generally a percentage of conductivity expressed as p.p.m. The percentage varies according to the instrument supplier between the ranges of 50 and 70% of conductivity.

Suppliers of electrical conductivity meters are reviewed in Appendix I.

The characteristics of a range of some general-purpose and research instruments available from various suppliers are tabulated in Table 5.3. From this list it should be possible for the chemist to select an instrument which best meets his requirements.

5.4 Dissolved oxygen

Most dissolved oxygen meters for laboratory or field use nowadays are of the portable hand-held battery-operated type. These are usually based on Clark electrodes or a silver anode and a platinum cathode separated from the test solution by a Teflon membrane. All operate in the range $0-20 \, \text{mg} \, l^{-1}$ dissolved oxygen with an accuracy ranging between ± 0.1 and $\pm 2\%$. Generally, temperature compensation can be made in the range 0–30 or 40°C (Table 5.4 and Appendix). Several of these instruments are available with fittings on the probe rendering them suitable for the measurement of biochemical oxygen demand (i.e. fitting for insertion in BOD bottles) and these will replace the bulkier instruments formerly used in this application. Automatic temperature compensation achieved by means of a built-in thermistor probe is now becoming a standard feature of these instruments.

5.5 Colour

The eye attempts to assess colour by three parameters. These are how light or dark the colour is, how strong or vivid and finally the hue or what the colour actually is. Tri-stimulus colorimetry allows these variables to be quantified and objectively assessed. By the choice of three suitable filter sets – red, green and blue – reflected or transmitted light can be analysed automatically and the resulting energies transformed into a variety of colour values. These values can be used to accurately monitor colour, colour variation and trends.

As the data is in numerical form it can be processed to allow for accurate reporting of colour trends or the use of statistical control techniques.

Table 5.4. *Dissolved oxygen meters*

Supplier	Model	Dissolved oxygen range ($mg\,l^{-1}\,O_2$)	Accuracy ($mg\,l^{-1}\,O_2$)	Temp. compensation (°0)	Flow velocity ($cm\,s^{-1}$)
Palintest	PT 125	0–20	0.2	0–30	30
EDT	EC 0291	0–20	0.1	Automatic −30 to 105	–
Jenway	9010	0–20	0.1	Automatic 0–40	–
	3410 (with DO_2 probe)				
	3420 (with DO_2 probe)				
	POM2	0–20	±2	Automatic 0–40	–
	9060	0–20	±2	0–40	–
	9070	0–20	±2	0–40	–
Orion	SL 9	0–20	–	–	–

The Trivecter CL 6000 colour measuring system is in use in several Water Authorities for the colour measurement of water and can measure Hazen units to an accuracy of within 1 unit, both off- and on-line and in the presence of suspended matter.

5.6 Elemental analysis

As well as determining individual anions and cations containing particular elements it is desirable to be able, in the case of particular elements, to be able to determine the total element content of the sample. Thus, in addition to nitrate, nitrite and ammonium it is frequently required to determine total nitrogen in the sample. In the case of halogens, for example, in addition to determining individual halogen-containing compounds, e.g. chloride, bromide and haloforms, it may be required to determine total halide or total organohalogen. In addition to water samples measurements of total element might be required on solid samples such as river or oceanic sediments.

Available commercial instrumentation for the determination of the following total element is the subject matter of this chapter:

1 halogen
2 sulphur
3 halogens and sulphur
4 nitrogen
5 carbon, hydrogen and nitrogen
6 nitrogen, carbon and sulphur
7 chemical oxygen demand
8 sample digestion procedures including COD

5.6.1 Halogens

Total halide

The Dohrmann DX 20B system is based on combustion of the sample to produce the hydrogen halide, which is then swept with a microcoulometric cell and estimated. It is applicable at total halide concentrations up to $1000 \, \mu g \, l^{-1}$ with a precision of $\pm 2\%$ at the $10 \, \mu g \, l^{-1}$ level. The detection limit is about $0.5 \, \mu g \, l^{-1}$. Analysis can be performed in 5 minutes. A sample boat is available for carrying out analysis of solid samples. The instrument has been applied to waste waters, soils and sediments.

Total organic halide

The Dohrmann DX-20A system is the DX 20B system with an additional module which makes it possible to measure total organic halides including chlorine, bromine and iodine in potable water. It features mini-column extraction of the sample with granulated activated carbon to preconcentrate organic halides at ultra-trace levels prior to combustion of the concentrate. An optional gas sparger attachment is available for determining purgeable organic chlorine compounds. Inorganic halides are removed from sample extracts by a nitrate wash so that only organic halides are reported. The technique is suitable for measurements of total organic halogen in potable water and drinking water by the Environmental Protection Agency method 8600, 9020 and 4501 and standard methods 506.

This instrument, therefore, has full capability for measurements in liquid or solid samples of total organic halogen (TOX) purgeable (volatile) organic halogen (POX) extractable organic halogen (EOX) and total halogen (TX).

Mitsubishi also supply a microprocessor-controlled automatic total halogen analyser (model TOX-10) (Figure 5.1(a)) which is very similar in operating principles to the Dohrmann instruments discussed above, i.e. combustion at 800–900°C followed by coulometric estimation of hydrogen halide produced.

Recent studies have indicated that total organic halogen contents of city potable water are 3 to 10 times greater than the volatile purgeable trihalomethane contents, making the need for total organic halogen content strong.

The TOX-10 can measure both purgeable and total organic halogen, some typical values being quoted in Table 5.5.

Recoveries of halogenated organics range from 92% (1,2 dibromoethane) to 105% (*m*-chlorobenzoic acid).

Table 5.5.

	Total organic halogen (TOX)	*Purgeable organic halogen (POX)*
City water A	86	10
City water B	130	35
City water C	150	48
City water D	230	25

Recoveries of halogenated organics range from 92% (1,2 dibromoethane) to 105% (*m*-chlorobenzoic acid)

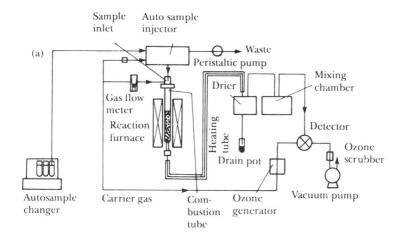

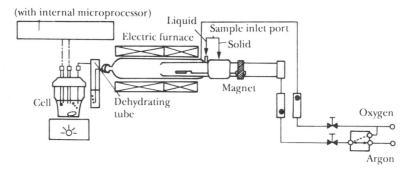

Figure 5.1 *Elemental analysis supplied by Mitsubishi (a) TN-05 nitrogen analyser; (b) TOX-10 total halogen analyser*

5.6.2 Sulphur

The Mitsubishi trace sulphur analyser models TS-02 and TN-02(S) is again a microcombustion procedure in which sulphur is oxidized to sulphur dioxide, which is then titrated coulometrically with triiodide ions generated from iodide ions:

$$SO_2 + I_3^- + H_2O \rightarrow SO_3 + 3I^1 + 2H^+$$
$$3I^1 \rightarrow I_3^- + 2e^-$$

5.6.3 Total sulphur/total halogen

The Mitsubishi TSX-10 halogen–sulphur analyser expands the technology of the TOX-10 to include total chlorine and total sulphur measurement. The model TSX-10, which consists of the TOX-10 analyser module and a sulphur detection cell, measures total sulphur and total chlorine in liquid and solid samples over a sensitivity range $mg\,l^{-1}$ to percent.

Dohrmann also produce an automated sulphur and chlorine analyser (models MCTS 130/120). This instrument is based on combustion microcoulometric technology.

5.6.4 Total bound nitrogen

Mitsubishi supply two total nitrogen analysers: the Model TN-10 and the model TN-05 microprocessor control chemiluminescence total nitrogen analysers (Figure 5.1(b)).

This instrument measures down to $\mu g\,l^{-1}$ amounts of nitrogen in solid and liquid samples.

The sample is introduced into the combustion tube packing containing oxidative catalyst under oxygen carrier gas. High-temperature oxidation (800–900°C) occurs and all chemically bound nitrogen is converted to nitric oxide (NO), R-N $\rightarrow$ CO_2 + NO. Nitric oxide then passes through a drier to remove water formed during combustion and moves to the chemiluminescence detector, where it is mixed with ozone to form excited nitrogen dioxide ($NO_2^\star$)

$$NO + O_3 \rightarrow NO_2^\star + O_2 \rightarrow NO_2 + O_2 + h_\upsilon$$

Rapid decay of the $NO_2^\star$ produces light in the 590–2900 nm range. It is detected and amplified by a photomultiplier tube. The result is calculated from the signal produced and printed out in milligrams per litre or as a percentage.

A wide 0.01 to $500\,mg\,l^{-1}$ detection range is possible for water samples. Coefficients of variation on water samples ranged from 0.88% at the $2.54\,mg\,l^{-1}$ level in river water to 3.1% at the $51\,mg\,l^{-1}$ level in sea water.

Dohrmann also supply an automated nitrogen analyser with video display and data processing (model DN-1000) based on similar principles which is applicable to the determination of down to $0.1\,mg\,l^{-1}$ nitrogen in solid and liquid samples.

Equipment for automated Kjeldahl determinations of organic nitrogen in water and solid samples is supplied by Tecator Ltd. Their Kjeltec system 1 streamlines the Kjeldahl procedure resulting in higher speed and accuracy compared to classical Kjeldahl measurements.

5.6.5 Carbon, hydrogen and nitrogen

Perkin-Elmer supply an analyser (model 2400 CHN) suitable for determining these elements in river and oceanic sediment samples and sewage sludges.

In this instrument the sample is first oxidized in a pure oxygen environment. The resulting combustion gases are then controlled to exact conditions of pressure, temperature and volume. Finally the product gases are separated under steady-state conditions and swept by helium or argon into a gas chromatograph for analysis of the components. The equipment is supplied with a 60 position autosampler and microprocessor controller covering all system functions, calculation of results and on-board diagnostics. Analysis time is 5 minutes.

5.6.6 Nitrogen, carbon and sulphur

The NA 1500 analyser supplied by Carlo Erba is capable of determining these elements in 3–9 minutes in amounts down to $10 \, mg \, l^{-1}$ with a reproducibility of ±0.1%. A 196 position autosampler is available.

'Flash combustion' of the sample in the combustion reactor is a key feature of the NA 1500. It results when the sample is dropped into the combustion reactor which has been enriched with pure oxygen. The normal temperature in the combustion tube is 1020°C and reaches 1700–1800°C during the flash combustion.

In the chromatographic column the combustion gases are separated so that they can be detected in sequence by the thermal conductivity detector (TCD). The TCD output signal is proportional to the concentration of the elements.

A data processor plots the chromatogram, automatically integrates the peak areas and prints retention times, percent areas, baseline drift and attenuation for each run. It also computes blank values, constant factors and relative average elemental contents.

5.6.7 Total organic carbon

Dohrmann supply a wide range of total organic carbon analysers characteristics of which are enumerated in Table 5.6.

The operating principle of these analysers involves a process whereby a persulphate reagent is continuously pumped at a low flow rate through the injection port (and the valve of the autosampler) and then into the UV reactor. A sample is acidified, sparged and injected directly into the reagent

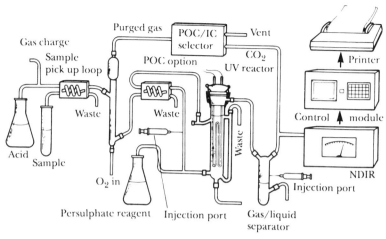

(a)

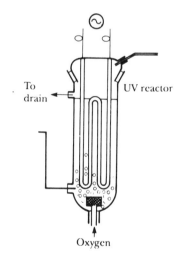

(b)

Figure 5.2 *Dohrmann DC-180 total organic carbon analyser (a) layout; (b) detail of UV reactor*

stream. The mixture flows through the reactor where organics are oxidized by the photon-activated reagent. The light-source envelope is in direct contact with the flowing liquid. Oxidation proceeds rapidly, the resultant carbon dioxide is stripped from the reactor liquid and carried to the carbon dioxide specific non-dispersive IR detector (NDIR).

Table 5.6. *Total organic carbon analysis as suplied by Dohrmann*

Model	Description	Sample types	Analysis time (min)	Principle*	Precision	Features	Detection limits
DC 85 A	Low-cost analyser	Soils, sediments, water	5	Combustion at 800°C in platinum boat, non-dispersive IR detector	±2% at 2 mg l^{-1}	Instrument expandable to include UV persulphate and vial persulphate oxidation, purgeables measurement	100 µg l^{-1}
DC-88	Low-cost analyser	Water	5	Persulphate oxidation at 100°C. Non-dispersive IR detector	100 mg l^{-1} or 2%, whichever greater	Instrument expandable to include UV persulphate oxidation and fast oxidation methods	100 µg l^{-1}
DC-80	Automated analyser	Water saline samples	3–4	Low-temperature promoted persulphate oxidation	±2% wt. 100–4000 mg l^{-1}	120-tube autosampler date integrity review. Purgeable organic carbon measurements can be made: Operating modes organic carbon external sparge total carbon inorganic carbon purgeable carbon estimate by difference purgeable carbon determined	–

DC-90	Second-generation high-temperature analyser	3 ground surface waters, lakes, oceans, potable water	High-temperature combustion in ceramic tube	From RSD of 0.6% at 6 mg l^{-1} to RSD of 0.5% at 107 mg l^{-1}	Unattended operation possible, loop-sampler option, autosampler printer option inorganic, carbon, purgeable carbon, total carbon and total organic carbon measurements possible	100 μg l^{-1}
LC-180	Automated analyser	Water, saline water, high-purity water, potable water	Combined UV persulphate oxidation (in combined cell)	±2% or 5 mg l^{-1} whichever greater	Highly automated inorganic carbon, non-purgeable organic carbon, purgeable organic carbon measurements possible. Automatic sample size selection according to type of sample, video display/microprocessor report on means and S.D. 45 sample autosampler	10 μg l^{-1}

* Inorganic carbon first removed by acidification and inert gas sparging.

As mentioned above, many variants of the Dohrmann total organic carbon analyser are available, ranging from low-cost non-automated analysers based on sample combustion in a platinum boat (DC8JA) or using persulphate oxidation/ultraviolet irradiation (DC 88) to top-of-range fully automated and computerized systems based on combustion in a ceramic tube (DC 90) or combined simultaneous persulphate–ultraviolet oxidation (DC 180). Only one of these systems, the DC 180, is discussed below in any detail.

As shown in Figure 5.2 sample transfer in the DC 180 is facilitated by gas pressure. Once the pick-up loop is filled a gas chase delivers the sample to the sparger. The DC 180 adds a preset amount of acid to the sample. Inorganic carbon is released in the form of carbon dioxide. Together with the purgeable organic carbon (POC) it is removed by sparging. The sample is now ready for non-purgeable organic carbon (NPOC) analysis.

Measuring non-purgeable organic carbon

A separate and independent injection loop dispenses the sample for non-purgeable organic carbon measurements.

In the reactor combined UV persulphate oxidation ensures quantitative total organic carbon recovery. The resulting carbon dioxide with entrained water goes through a gas/liquid separator, a water trap and drier before it enters the non-dispersive infrared analyser detector where the evolved carbon dioxide is measured.

Measuring inorganic carbon

If it is required to quantify inorganic carbon the sparged gas may be directed to the non-dispersive infrared analyser for quantification.

Measuring purgeable organic carbon

The volatile fraction from the sparger contains both carbon dioxide and purgeable organic carbon. If purgeable organic carbon measurements are required, carbon dioxide is removed by the lithium hydroxide scrubber. There are two options for oxidizing purgeable organic carbon. The UV persulphate reactor will convert most purgeable organic carbon except for fully halogenated organics such as Freons and carbon tetrachloride. In the case of such organics, the high-temperature reactor will be required.

Measuring purgeable organic carbon inorganic carbon and non-purgeable organic carbon

In addition to giving enhanced sensitivity and greater recovery for the full range of purgeable organic carbons, the purgeable organic carbon accessory permits the analysis of purgeable organic carbon, inorganic carbon and non-purgeable organic carbon on one sample. The DC 180 reports all three of these parameters plus total organic carbon as a sum of non-purgeable organic carbon and purgeable organic carbon and total carbon as a sum of all three parameters.

Measuring total organic carbon

The DC 180 will calculate total organic carbon based on purgeable organic carbon and non-purgeable organic carbon results and include it in the report. Alternatively, total organic carbon may be determined as the difference of total carbon less inorganic carbon.

Shimadzu TOC-500 total organic carbon analyser

This is a fully automated system capable of determining between $1\,\mu g\,l^{-1}$ and $3000\,\mu g\,l^{-1}$ total organic carbon. It is applicable to saline and non-saline waters and is equipped with a 36-place autosampler, microprocessor and printer. Total organic carbon measurements down to $40\,\mu g\,l^{-1}$ in ultra-pure water have been achieved at a coefficient of variation of 16.3%.

OIC Analytical instruments produce the fully computerized model 700 total organic carbon analyser. This is applicable to potable water, ground water, waste water, sea water, ultra-pure water, cooling water, soils and sediments. Persulphate oxidation at 90–100°C non-dispersive infrared spectroscopy is the principle of this instrument. It has the ability to measure total organic carbon, total inorganic carbon and purgeable organic carbon in the same sample. The precision is $\pm 2.0\,\mu g\,l^{-1}$ carbon.

5.6.8 Acid digestion systems

In water analysis it is frequently required to carry out acid (and occasionally other reagent) digestions on larger batches of samples preparatory to carrying out an analytical finish. Examples include acid digestions of water or sediments prior to the determination of metals, Kjeldahl determinations of nitrogen and determinations of chemical oxygen demand.

Tecator supply multiple digestion systems for Kjeldahl nitrogen (System 6/20, 1030, 12/40) and chemical oxygen demand. Skalar supply their systems 500 which will house 20×250 ml or 42×75 ml sample digestion tubes and is very useful for digestions prior to the determination of metals, chemical oxygen demand, etc.

6 *Organic substances*

The identification and determination of traces of organic substances in water samples is a subject that has made tremendous advances in recent years. The demands made on water chemists in terms of specificity and sensitivity in carrying out these analyses have become greater and greater with the increasing realization that organic substances from industrial sources are permeating the ecosystem and identification and measurements of minute traces of these are required in potable, river and ground waters and even in rainwater. At the same time, measurements in industrial effluent outfalls are necessary in order to control the rate of release of these substances.

For the more volatile components of water samples, i.e. those with boiling points up to about 250°C, gas chromatography has been a favoured technique for several decades. However, with the realization that retention time measurements alone are insufficient to identify organics there has been an increasing move in recent years to connect a gas chromatograph to a mass spectrometer in order to provide unequivocal identifications. Element-specific detectors are another recent development.

For even the most volatile substances such as trihalomethanes and halogenated aliphatic compounds, commercial instrumentation is now available for headspace analysis and purge and trap techniques, whereby the gas space above a sample or the gas pumped through a sample is swept into a gas chromatograph for analysis.

A limitation of gas chromatography is that it cannot handle less volatile components of water samples and these comprise a high proportion of the total organics content of the sample. For this reason increasing attention is being paid to the application of high-performance liquid chromatography in water analysis. Again, when positive identifications are required, a mass spectrometer is connected to the outlet of the chromatograph.

Supercritical fluid chromatography is a recent development which may find increasing use in water analysis.

There are also more specific chromatographic techniques for which instrumentation is now available, such as amino acid analysis, ion-exchange chromatography and ion chromatography.

6.1 Gas chromatography

The water chemist will certainly require what is referred to as a high-performance instrument. If it is his intention to couple the gas chromatograph to a mass spectrometer, then he should consider the purchase of the gas chromatograph and mass spectrometer as a single purchase, so as to obtain a well-married system (discussed further in Section 6.6).

The basic requirements required of a high-performance gas chromatograph are as follows:

- Sample is introduced to the column in an ideal state, i.e. uncontaminated by septum bleed or previous sample components, without modification due to distillation effects in the needle and quantitatively, i.e. without hold-up or adsorption prior to the column.
- The instrument parameters that influence the chromatographic separation are precisely controlled.
- Sample components do not escape detection, i.e. highly sensitive, reproducible detection and subsequent data processing are essential.

There are two types of separation column used in gas chromatography, capillary column and packed column.

Packed columns are still used extensively, especially in routine analysis. They are essential when sample components have high partition coefficients and/or high concentrations. Capillary columns provide a high number of theoretical plates, hence a very high resolution, but they cannot be used in all applications because there are not many types of chemically bonded capillary columns. Combined use of packed columns of different polarities often provides better separation than with a capillary column. It sometimes happens that a capillary column is used as a supplement in the packed-column gas chromatograph. It is best, therefore, to house the capillary and packed columns in the same column oven and use them selectively. In the screening of some types of samples, the packed column is used routinely and the capillary column is used when more detailed information is required.

Conventionally, it is necessary to use a dual column flow line in packed-column gas chromatography to provide sample and reference gas flows. The recently developed electronic base line drift compensation system allows a simple column flow line to be used reliably.

Recent advances in capillary column technology presume stringent performance levels for the other components of a gas chromatograph as column performance is only as good as that of the rest of the system. One of the most important factors in capillary column gas chromatography is that a high repeatability of retention times be ensured even under adverse ambient conditions.

These features combine to provide ±0.01 minute repeatability for peaks having retention times as long as 2 hours (other factors being equal).

Another important factor for reliable capillary column gas chromatography is the sample injection method. Various types of sample injection points are available. The split/splitless sample injection port unit series is designed so that the glass insert is easily replaced and the septum is continuously purged during operation. This type of sample injection unit is quite effective for the analysis of samples having high-boiling point compounds as the major components.

In capillary column gas chromatography, it is often required to raise and lower the column temperature very rapidly and to raise the sample injection port temperature. In one design of gas chromatograph, the Shimadzu GC 14-A, the computer-controlled flap operates to bring in the external air to cool the column oven rapidly – only 6 minutes from 500°C to 100°C. This computer-controlled flap also ensures highly stable column temperature when it is set to a near-ambient point. The lowest controllable column temperature is about 26°C when the ambient temperature is 20°C.

6.1.1 Instrumentation

Shimadzu gas chromatographs

This is a typical high-performance gas chromatograph version (see Table 6.1 for further details).

The inner chamber of the oven has curved walls for smooth circulation of air; the radiant heat from the sample injection port units and the detector oven is completely isolated. These factors combine to provide demonstrably uniform temperature distribution. (The temperature variance in a column coiled in a diameter of 20 cm is less than $\pm 0.75\,°K$ at a column temperature of 250°C).

When the column temperature is set to a near ambient temperature, external air is brought into the oven via a computer-controlled flap, providing rigid temperature control stability. (The lowest controllable column temperature is 24°C when the ambient temperature is 18°C and the injection part temperature is 250°C. The temperature fluctuation is less than $\pm 0.1\,°K$ even when the column temperature is set at 50°C.

This instrument features five detectors (Table 6.1). In the flame ionization detector, the high-speed electrometer, which ensures a very low noise level, is best suited to trace analysis and fast analysis using a capillary column.

Samples are never decomposed in the jet, which is made of quartz.

Carrier gas, hydrogen, air and make-up gas are separately flow-controlled. Flow rates are read from the pressure flow-rate curves.

In the satellite system, one or more satellite gas chromatographs (GC-14 series) are controlled by a core gas chromatograph (e.g. GC 16A series). Since the control is made externally, the satellite gas chromatographs are not required to have control functions (the keyboard unit is not necessary).

When a GC 16A series gas chromatograph is used as the core, various laboratory-automation-oriented attachments such as bar-code reader and a

Table 6.1. *Commercial gas chromatographs*

Manu-facturer	Model	Packed column	Capillary column	Detectors	Sample injection point system	Keyboard control	Link to computer	Visual display	Printer	Core instrument amenable to tap automation	Temperature programming/ isothermal	Cryogenic unit (sub-ambient chroma-tography)
Shimadzu	GC-14A	Yes	Yes	FID ECD FTD FPD TCD (all supplied)	1. Split–splitters 2. Glass insert for single column 3. Glass insert to dual column 4. Cool on column system unit 5. Moving needle system 6. Rapidly ascending temperature vaporizer	Yes	Yes	No	No	No	Yes/Yes	No
Shimadzu	GC 15A	Yes	Yes	FID ECD FTD FPD TCD	1. Split–splitters 2. Direct sample injection (capillary column) 3. Standard sample injector (packed column) 4. Moving precolumn system (capillary columns) 5. On column (capillary columns)	No	Yes	Yes	Yes	No	Yes/Yes	No
Shimadzu	GC 16A		Yes	FID FCD FTD FPD TCD (all supplied)	Split–splitters	Yes	Yes	Yes	Yes	No	Yes/Yes	No
Shimadzu	GC 8A	Yes	Yes	FID FCD FPD TCD Single detector instruments (detector chosen on purchase)	1. Point for packed columns 2. Point for capillary columns 3. Split–splitters	Yes Not built in	Optional	No	Optional	No	Yes temp prgramming GC 8APT (TCD detector) GC 8APF (FID detector) GV 8APFD (FID detector) isothermal: GC 8AIT (TCD detector) GC 8A1F (FID detector) GC 8A1E (ECD detector)	No

Manufacturer	Model			Detector systems	Injection systems							
Perkin-Elmer	8410	Yes	Yes	Single detector instrument (detector chosen on purchase) FID ECD FTD FPD TCD	1. Flash vaporization 2. Slit–splitless injector 3. Manual or automatic gas sampling valves 4. Manual or automatic liquid sampling valves	No	No	No	No	No	Yes/Yes	Yes, down to −80°C
Perkin-Elmer	8420	No	Yes	Single detector instrument (chosen on purchase) FID ECD FTD FPD TCD	1. Programmable temperature vaporizer 2. Split–splitless injector 3. Direct on column injector	No	No	No	No	No	Yes/Yes	Yes, down to −80°C
Perkin-Elmer	8400 and 8500	Yes	Yes	Dual detector instrument (detectors chosen from following) FID ECD FTD FPD TCD	Can be fitted with any combination of above injection systems	Yes	Yes	Yes	Yes (GP100 printer plotter)	Yes	Yes/Yes	Yes, down to −80°C
Perkin-Elmer	8700	Yes	Yes	FID ECD FPD TCD Hall E.C. photoionization dual detector instrument. (Detectors chosen from above list)	1. Flash vaporization 2. Split–splitless 3. Programmable temperature vaporizer 4. Gas sampling valve 5. Liquid sampling valve	Yes	Yes	Yes	Yes	–	Yes/Yes	Yes
Nordion	Micromat HRGC 412	No	Yes	Dual simultaneous detector combinations from the following: FID ECD FTD Photoionization Hall E.C.	1. Split–splitless 2. On-column injector	Yes	Yes	Yes	Yes	–	Yes/Yes	No
Siemens	SiChromat 1–4 (single oven) SiChromat 2–8 (dual oven) for multidimensional GC)	Yes	Yes	FID ECD FTD FPD TCD Helium detector	1. Liquid–liquid packed columns 2. Split–splitless 3. Temperature programmable 4. On-column 5. Liquid injector valve on-line 6. Gas injection valve 7. Rotary as injection valve	No	No	Yes	Yes	–	Yes/Yes	No

magnetic-card reader become compatible: a labour-saving system can be built, in which the best operational parameters are automatically set. Each satellite gas chromatograph (GC 14A series) operates as an independent instrument when a keyboard unit is connected.

The IC card operated gas chromatography system consists of a GC-14A series gas chromatograph and a C-R5A Chromatopac data processor. All of the chromatographic and data processing parameters are automatically set simply by inserting the particular IC card. This system is very convenient when one GC system is used for the routine analysis of several different types of samples.

One of the popular trends in laboratory automation is to arrange for a personal computer to control the gas chromatograph and to receive data from the GC to be processed as desired. Bilateral communication is made via the RS-232C interface built in a GC 14A series gas chromatograph. A system can be built to meet requirements.

A multidimensional gas chromatography system (multi-stage column system) is effective for analysis of difficult samples and can be built up by connecting several column ovens, i.e. tandem GC systems, each of which has independent control functions such as for temperature programming.

The Shimadzu GC 15A and GC 16A systems are designed not only as independent high-performance gas chromatographs but also as core instruments (see above) for multi-gas-chromatography systems (i.e. several gas chromatographs in the laboratory linked to a central management system) or computerized laboratory automation systems. The GC 16A has a keyboard, the GC 15A does not. Other details of these instruments are given in Table 6.1. The Shimadzu GC 8A range of instruments do not have a range of built-in detectors but are ordered either as temperature programmed instruments with TCD, FID or FPD detectors or as isothermal instruments with TCD, FID or ECD detectors (Table 6.1).

Perkin-Elmer supply a range of instruments including the basic models 8410 for packed and capillary work and the 8420 for dedicated capillary work, both supplied on purchase with one of the six different types of detection (Table 6.1). The models 8400 and 8500 are more sophisticated capillary column instruments capable of dual detection operation with the additional features of keyboard operation. Screen graphics method storage, host computer links, data handling and compatibility with laboratory automation systems. Perkin-Elmer supply a range of accessories for these instruments including an autosampler (AS-8300) an infrared spectrometer interface, an automatic headspace accessory (HS101 and H5-6) an autoinjector device (AI-I) also a catalytic reactor and a pyroprobe (CDS 190) and automatic thermal disorption system (ATD-50) (both useful for examination of sediments).

The Perkin-Elmer 8700, in addition to the features of the models 8400 and 8500, has the ability to perform multi-dimensional gas chromatography.

The optimum conditions for capillary chromatography of material heart cut from a packed column demand a highly sophisticated programming system. The software provided with the model 8700 provides this, allowing methods to be linked so that pre-column and analytical column separations are performed under optimum conditions. Following the first run, in which components are transferred from the pre-column to the on-line cold trap, the system will reset to a second method and, on becoming ready, the cold trap is desorbed and the analytical run automatically started.

Other applications of the model 8700 system include fore-flushing and back-flushing of the pre-column, either separately or in combination with heart cutting, all carried out with complete automation by the standard instrument software.

There are many other suppliers of gas chromatography equipment, some of which are discussed further in Table 6.1.

6.1.2 Applications

Gas chromatography has been employed for the determination of a wide range of organics in water and sediments and other environmental samples such as fish.

Figure 6.1(a) shows a gas chromatogram obtained using an electron capture detector in the determination in water of materials as variable as organic halogens and chlorinated insecticides. Figure 6.1(b), on the other hand, shows a dual chromatogram indicating the presence of a very complex mixture of chlorinated compounds present in the muscle of trout which had been exposed to the effluent from a pulp mill.

6.2 Headspace samplers

Headspace analysis is a method of choice for the determination of volatile compounds in heterogeneous multi-component water samples. The classic application of this technique is perhaps the determination of trihalomethanes in potable water, but numerous other applications have been found such as the determination of chlorinated aliphatics and volatile hydrocarbons in water. The principle of the analysis is quite simple. The sample is placed in a container leaving a large headspace, which is filled with an inert gas (sometimes under pressure) which also serves as the gas chromatograph carrier gas.

Several gas chromatograph manufacturers (Table 6.2) now produce apparatus for headspace analysis. In the past, headspace sampling has often been mentioned in conjunction with low reproducibility. By adding a

138

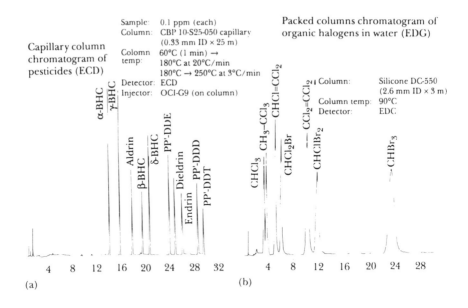

Capillary column chromatogram of pesticides (ECD)

Packed columns chromatogram of organic halogens in water (EDG)

Sample: 0.1 ppm (each)
Column: CBP 10-S25-050 capillary (0.33 mm ID × 25 m)
Colomn temp: 60°C (1 min) →
180°C at 20°C/min
180°C → 250°C at 3°C/min
Detector: ECD
Injector: OCI-G9 (on column)

Column: Silicone DC-550 (2.6 mm ID × 3 m)
Column temp: 90°C
Detector: EDC

(a)

(b)

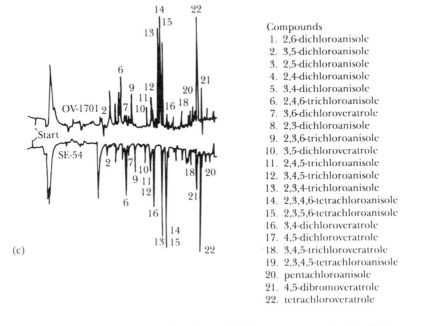

OV-1701

Start

SE-54

(c)

Compounds
1. 2,6-dichloroanisole
2. 3,5-dichloroanisole
3. 2,5-dichloroanisole
4. 2,4-dichloroanisole
5. 3,4-dichloroanisole
6. 2,4,6-trichloroanisole
7. 3,6-dichloroveratrole
8. 2,3-dichloroanisole
9. 2,3,6-trichloroanisole
10. 3,5-dichloroveratrole
11. 2,4,5-trichloroanisole
12. 3,4,5-trichloroanisole
13. 2,3,4-trichloroanisole
14. 2,3,4,6-tetrachloroanisole
15. 2,3,5,6-tetrachloroanisole
16. 3,4-dichloroveratrole
17. 4,5-dichloroveratrole
18. 3,4,5-trichloroveratrole
19. 2,3,4,5-tetrachloroanisole
20. pentachloroanisole
21. 4,5-dibromoveratrole
22. tetrachloroveratrole

Figure 6.1 *Gas chromatography of organic halogen compounds: (a), (b) in water; (c) in trout exposed to kraft pulp mill effluent*

gas-tight syringe and by employing the latest instrumental refinements to the proven pressure-balanced sampling method, better reproducibility is obtained than that obtained with liquid injection. Typical performance is now better than ±1% relative standard deviation.

Under the prevailing equilibrium conditions a proportion of the volatiles in the sample transfer to the gas-filled headspace, which is then withdrawn and analysed by gas chromatography.

6.2.1 Instrumentation

The HS-100/HS-101 automatic headspace analysers produced by Perkin-Elmer employ a pneumatic pressure-balanced system. The HS-100 model is suitable for use with Perkin-Elmer Sigma 2000 series of gas chromatograph whilst the HS-101 is designed for use with the 8000 series. With the advanced microprocessor system of the 8000 series and the moving-needle design the HS-101 offers new and important application possibilities unavailable with other headspace injection systems. The HS-101 has a 100-sample storage magazine and is suitable for unattended and night operation.

Suppliers of headspace analysers are reviewed in Table 6.2.

6.2.2 Applications

In Figure 6.2 are shown a gas chromatogram obtained for (a) a mixture of volatile halogenated hydrocarbons and (b) gasoline in water obtained respectively using a packed and a capillary gas chromatographic column.

6.3 Purge and trap gas chromatography

This is an alternative technique to headspace analysis for the identification and determination of volatile organic compounds in water. The sample is purged with an inert gas for a fixed period of time. Volatile compounds are sparged from the sample and collected on a solid sorbent trap – usually activated carbon. The trap is then rapidly heated and the compounds collected and transferred as a plug under a reversed flow of inert gas to an external gas chromatograph. Chromatographic techniques are then used to quantify and identify sample components.

6.3.1 Instrumentation

OIC Analytical Instrument (Appendix) supply the 4460 A purge and trap concentrator. This is a microprocessor-based instrument with capillary

Table 6.2. *Headspace samplers*

Supplier	Headspace analyser model no.	Capillary/ packed	Compatible gas chromatograph model no.	Automated	Microprocessor control	Multiple headspace analysis option	Pressure option	Sample carousel	Isothermal/ temp programming at 9°C	Thermostatic sample temperature
Perkin-Elmer	HS-101	Capillary	8000 series	Yes	Yes	Yes	Yes	100 sample	Yes/Yes	35–150
Perkin-Elmer	HS-100	Capillary	Sigma 2000	Yes	Yes	Yes	Yes	600 sample	Yes/Yes	35–150
Perkin-Elmer	HS-6	Packed or capillary	Sigma 2000 models 3920, 900, 990, F-22 8-30	No	No	Yes	Yes	No	–	40–196
								No		190
Sievers	Headspace	Packed or capillary	Sichromat 1–4 Sichromat 2–8	No	No	–	–	No	Yes/Yes	
Shimadzu	HSS-2A	Packed or capillary	GC-9A	Yes	Yes	No	No	40 sample	Yes/Yes	40–150

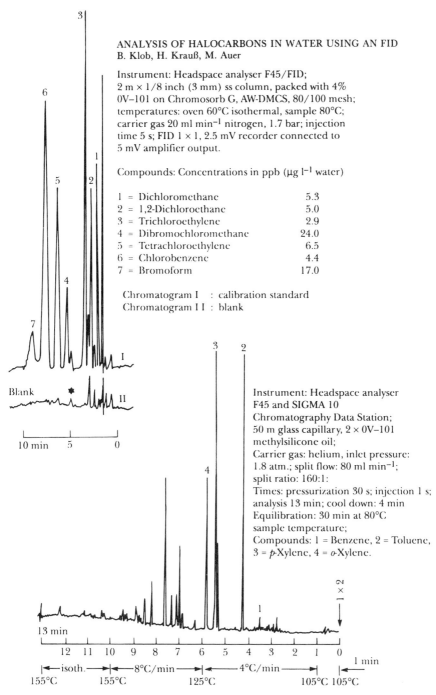

ANALYSIS OF HALOCARBONS IN WATER USING AN FID
B. Klob, H. Krauß, M. Auer

Instrument: Headspace analyser F45/FID;
2 m × 1/8 inch (3 mm) ss column, packed with 4%
OV–101 on Chromosorb G, AW-DMCS, 80/100 mesh;
temperatures: oven 60°C isothermal, sample 80°C;
carrier gas 20 ml min⁻¹ nitrogen, 1.7 bar; injection
time 5 s; FID 1 × 1, 2.5 mV recorder connected to
5 mV amplifier output.

Compounds: Concentrations in ppb (μg l⁻¹ water)

1 = Dichloromethane	5.3
2 = 1,2-Dichloroethane	5.0
3 = Trichloroethylene	2.9
4 = Dibromochloromethane	24.0
5 = Tetrachloroethylene	6.5
6 = Chlorobenzene	4.4
7 = Bromoform	17.0

Chromatogram I : calibration standard
Chromatogram II : blank

Instrument: Headspace analyser
F45 and SIGMA 10
Chromatography Data Station;
50 m glass capillary, 2 × OV–101
methylsilicone oil;
Carrier gas: helium, inlet pressure:
1.8 atm.; split flow: 80 ml min⁻¹;
split ratio: 160:1:
Times: pressurization 30 s; injection 1 s;
analysis 13 min; cool down: 4 min
Equilibration: 30 min at 80°C
sample temperature;
Compounds: 1 = Benzene, 2 = Toluene,
3 = p-Xylene, 4 = o-Xylene.

Figure 6.2 *Headspace analysis: (a) halocarbons; (b) gasoline in water*

column capability. It is supplied with an autosampler capable of handling 76 sample vials. Two automatic rinses of sample lines and vessel purge are carried out between sample analyses to minimize carry-over.

Tekmar are another supplier of purge and trap analysis equipment. Their LSC 2000 purge and trap concentrator features glass-lined stainless steel tubing, a menu-driven programming with four-method storage and a cryofocusing accessory.

Cryofocusing is a technique in which only a short section of the column or a pre-column is cooled. In its simplest form a section of the column near the inlet is immersed in a flask of coolant during desorb. After desorb the coolant is removed and the column allowed to return to the oven temperature.

Performance aspects of volatiles organics analysis by purge and trap capillary column gas chromatography with flame ionization detectors has been discussed by Westendorf (1986).

6.4 High-performance liquid chromatography

Modern high-performance liquid chromatography has been developed to a very high level of performance by the introduction of selective stationary phases of small particle sizes, resulting in efficient columns with large plate numbers per litre.

Separation columns

There are several types of chromatographic columns used in high-performance liquid chromatography:

(a) Reversed phase chromatography
The most commonly used chromatographic mode in HPLC is reversed-phase chromatography. Reversed-phase chromatography is used for the analysis of a wide range of neutral and polar organic compounds. Most common reversed phase chromatography is performed using bonded silica-based columns, thus inherently limiting the operating pH range to 2.0–7.5. The wide pH range (0–14) of some columns (e.g. Dionex Ion Pac NSI and NS 1–5 μ columns) removes this limitation, consequently they are ideally suited for ion-pairing and ion-suppression reversed-phase chromatography: the two techniques which have helped extend reverse-phase chromatography to ionizable compounds.

High-sensitivity detection of non-chromophoric organic ions can be achieved by combining the power of suppressed conductivity detection with these columns. Suppressed conductivity is usually a superior approach to using refractive index or low UV wavelength detection.

(b) Reversed-phase ion-pairing chromatography

Typically, reversed-phase ion-pairing chromatography is carried out using the same stationary phase as reversed-phase chromatography. A hydrophobic ion of opposite charge to the solute of interest is added to the mobile phase. Samples which are determined by reversed-phase ion-packing chromatography are ionic and thus capable of forming an ion pair with the added counter ion. This form of reversed-phase chromatography can be used for anion and cation separations and for the separation of surfactants and other ionic types of organic molecules. An unfortunate drawback to using silica-based columns is that ion-pairing reagents increase the solubility of silica in water, leading to loss of bead integrity and drastically reducing column life. Some manufacturers (e.g. Dionex) employ neutral macroporous resins, instead of silica, in an attempt to widen the usable pH range and eliminate the effect of ion-pairing reagents.

(c) Ion-suppression chromatography

Ion suppression is a technique used to suppress the ionization of compounds (such as carboxylic acids) so they will be retained exclusively by the reversed-phase retention mechanism and chromatographed as the neutral species.

Column packings with an extended pH range are needed for this application as strong acids or alkalis are used to suppress ionization. In addition to carboxylic acids, the ionization of amines can be suppressed by the addition of a base to the mobile phase, thus allowing chromatography of the neutral amine.

(d) Ion-exclusion chromatography

Unlike the pellicular packings used for ion exchange, the packings used in ion exclusion are derived from totally sulphonated polymeric materials. Separation is dependent upon three different mechanisms:

- Donnan exclusion
- steric exclusion
- adsorption/partitioning

Donnan exclusion causes strong acids to elute in the void volumes of the column. Weak acids which are partially ionized in the eluent are not subject to Donnan exclusion and can penetrate into the pores of the packing. Separation is accomplished by differences in acid strength, size and hydrophobicity. The major advantage of ion exclusion lies in the ability to easily handle samples that contain both weak and strong acids. A good example of the power of ion exclusion is the routine determination of organic acids in sea water. Without ion exclusion, the high chloride ion concentration would present a serious interference.

Ion-exchange chromatography

Ion-exchange chromatography is based upon the differential affinity of ions for the stationary phase. The rate of migration of the ion through the column is directly dependent upon the type and concentration of ions that comprise the eluent. Ions with low or moderate affinities for the packing generally prove to be the best eluents. Examples are hydroxide and carbonate eluents for anion separations.

The stationary phases commonly used in HPLC are typically derived from silica substrates. The instability of silica outside the pH range 2 to 7.5 represents one of the main reasons why ion exchange separations have not been extensively used in HPLC. To overcome this, some manufacturers (e.g. Dionex) Ion-Pac columns supply a packing which is derived from crosslinked polystyrene which is stable throughout the entire pH range. This pH stability allows eluents of extreme pH values to be used so that weak acids such as carbohydrates (and bases) can be ionized.

6.4.1 Instrumentation

Elution systems

Four basic types of elution system are used in HPLC. This is illustrated below by the systems offered by LKB, Sweden:

(a) The isocratic system (Figure 6.3(a))
This consists of a solvent delivery for isocratic reversed phase and gel filtration chromatography.

This isocratic system provides an economic first step into high-performance liquid chromatography techniques. The system is built around a high-performance, dual-piston, pulse-free pump providing precision flow from 0.01 to 5 ml/min.

Any of the following detectors can be used with this system:

- fixed wavelength ultraviolet detector (LKB Unicord 2510)
- variable UV visible (190–600 nm)
- wavelength monitor (LKB 2151)
- rapid diode array spectral detector (LKB 2140) (discussed later)
- refractive index detector (LKB 2142)
- electrochemical detector (LKB 2143)
- wavescan EG software (LKB 2146)

(b) Basic gradient system (Figure 6.3(b))
This is a simple upgrade of the isocratic system with the facility for gradient elution techniques and greater functionality. The basic system provides for manual operating gradient techniques such as reversed-phase, ion-exchange

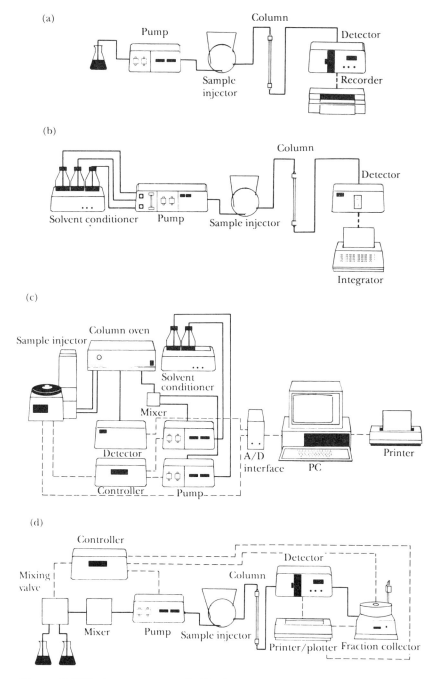

Figure 6.3 *Elution systems supplied by LKB, Sweden: (a) isocratic bioseparation system; (b) basic system; (c) advanced chromatography system; (d) inert system*

and hydrophobic interaction chromatography. Any of the detectors listed above under the isocratic system can be used.

(c) Advanced gradient system (Figure 6.3(c))

For optimum functionality in automated systems designed primarily for reversed-phase chromatography and other gradient techniques, the LKB advanced-gradient system is recommended. Key features include the following:

- a configuration that provides the highest possible reproducibility of results
- a two-pump system for highly precise and accurate gradient formation for separation of complex samples
- full system control and advanced method development provided from a liquid chromatography controller
- precise and accurate flows ranging from 0.01 to 5 ml/min

This system is ideal for automatic method for development and gradient optimization.

(d) The inert system (Figure 6.3(d))

By a combination of the use of inert materials (glass, titanium, and inert polymers) this system offers totally inert fluidics. Primary features of the system include the following:

- the ability to perform isocratic or gradient elution by manual means
- full system control from a liquid chromatography controller
- precise and accurate flows from $0.01–5\, ml\, min^{-1}$

This is the method of choice when corrosive buffers, e.g. those containing chloride or aggressive solvents, are used.

Chromatographic detectors

Details concerning the types of detectors used in high-performance liquid chromatography are given in Table 6.3. The most commonly used detectors are those based on spectrophotometry in the region 185–400 nm, visible ultraviolet spectroscopy with region 185–900 nm, post-column derivativiza-tion with fluorescence detection (see below), conductivity and those based on the relatively new technique of multiple wavelength ultraviolet detectors using a diode array system detector (see below). Other types of detectors available are those based on electrochemical principles, refractive index, differential viscosity and mass detection.

Post-column derivatization – fluorescence detectors

Modern column liquid chromatography has been developed to a very high level by the introduction of selective stationary phases of small particle sizes, resulting in efficient columns with large plate numbers per metre. The development of HPLC equipment has been built upon the achievements in column technology, but the weakest part is still the detection system. UV/vis and fluorescence detectors offer tremendous possibilities, but because of their specificity it is possible to detect components only at very low concentrations with a specific chromophore or fluorophore. The lack of a sensitive all-purpose detector in liquid chromatography like the flame ionization detector in gas chromatography, is still disadvantageous for liquid chromatography for the detection of important groups of compounds, such as amino acids. Consequently, chemical methods are increasingly used to enhance selectivity and sensitivity of detection. On-line post column derivatization started with the classic work of Spackmann *et al.* (1958) with amino acid analysis and has recently found increasing interest and use (Frei and Lawrence 1981a,b; Krull 1986; Engelhardt 1979; Engelhardt and Neue 1982; Engelhardt and Lillig 1985, 1986; Engelhardt *et al* 1985; Uihlein and Schwab 1982).

With on-line post-column detection the complexity of the chromatographic equipment increases. An additional pump is required for the pulseless and constant delivery of the reagent.

Diode array detectors

With the aid of a high-resolution ultraviolet diode array detector, the eluting components in a chromatogram can be characterized on the basis of their UV spectra. The detector features high spectral resolution (comparable to that of a high-performance UV spectrophotometer) and high spectral sensitivity. The high spectral sensitivity permits the identification of spectra near the detection limit, i.e. within the submilliabsorbance range, while complete spectral information can be obtained. Digital resolution of the spectra permits numerical processing and thus facilitates liquid chromatography/ultraviolet coupling.

In HPLC applications, such as environmental analysis, it is important that peak identification be carried out as fast and as distinctly as possible and this technique has been found to be the most efficient and economical alternative for peak characterization. Ultraviolet spectra do not have the same distinct characteristics as, for example, IR or mass spectra, yet by numerical processing, the spectra can be used for substance identification. The better the spectral and digital resolution of the UV spectra to be evaluated, the sooner numerical processing can be applied and the easier will it be to exploit the usually marginal differences in the form and position of spectral bands. With these aspects in mind several manufacturers (Varian, Perkin-Elmer, LKB and Hewlett Packard, see Table 6.3) have developed diode array systems. In the polychromator incorporated in the Perkin-Elmer LC 480 diode array system the light beam is dispersed within the range 190–430 nm

Table 6.3. *Detectors used in HPLC*

Type of detector		Supplier	Detection part no.	HPLC instrument part no.
Spectrophotometric (variable wavelength)	190–390 nm	Perkin–Elmer	LC-90	–
	195–350 nm	Kontron	735 LC	Series 400
	195–350 nm	Shimadzu	SPD-7A	LC-7A
	195–350 nm	Shimadzu	SPD-6A	LC-8A
	195–350 nm	Shimadzu	SPD-6A	LC-6A
	206–405 nm (fixed wavelength choice of 7 wavelengths between 206 and 405 nm)	LKB	2510 Uvicord SD	–
	190–370 nm ⎫	Cecil Instruments	Model 1937	Chrom-A-Scope
	190–400 nm ⎭		CE 1220	Series 1000
Variable wavelength UV–visible	190–600 nm	Varian	2550	2500
	190–600 nm	LKB	2151	Uvicord SD
	190–700 nm	Kontron	432	Series 400
	190–800 nm	Kontron	430	Series 400
	185–900 nm	Kontron	720 LC	Series 400
	200–570 nm	Kontron	740 LC	Series 400
	190–800 nm	Dionex	VDM II	Series 400
	190–750 nm	Isco	V4 ⎫	Microbo system
	214–660 nm (18 preset wavelengths)	Isco	UAS and 228 ⎭	
	195–700 nm	Shimadzu	SPD 7A	LC-7A
	195–700 nm	Shimadzu	SPD-6AV	LC-8A
	195–700 nm	Shimadzu	SPD-6AV	LC-6A

Detector type	Manufacturer	Wavelength	Programmably variable wavelength detector	
	Hewlett Packard	190–600 nm		9050 series
	Cecil Instruments	380–600 nm	CE 1200	Series 1000
	Applied Chromatography systems	190–800 nm	750/16 and 5750/11	–
Conductivity	Dionex	–	CDM 11	4500 i
	Roth Scientific		–	Chrom-A-scope
Electrochemical detector	Dionex	–	PAD-11	4500i
	LKB		2143	Wave-scan EG
	Roth Scientific		–	Chrom-A-scope
	Cecil Instruments		CE 1500	–
	PSA Inc.		5100A	–
	Applied Chromatography systems		650/350/06	–
Refractive index detector	LKB		2142	Wavescan E.G.
	Roth Scientific		–	Chrom-A-Scope
	Cecil Instruments		CE 1400	Series 1000
Differential viscosity mass detection (evaporative)	Roth Scientific		–	Chrom-A-Scope
	Applied Chromatography systems		750/14	–
Diode array	Varian		9060	2000L and 5001 5500 series
	Perkin-Elmer		LC135, LC 235 and LC 480	–
	LKB		2140	–
	Hewlett Packard		Multiple wavelength detector	1050 series

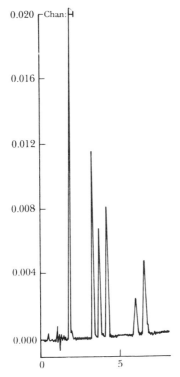

Figure 6.4 *Chromatogram of six polycyclic aromatic hydrocarbons, fluoranthene, benzo(b)fluoranthene, benzo(k)fluoranthene, benzo(a)pyrene, benzo(g, h, i)pery-lene and indeno(1,2,3-co)pyrene (injection volume: 10 µl)*

onto a diode array consisting of 240 light-sensitive elements. This effects a digital resolution of 1 nm, which thus satisfies the spectral resolution determined by the entrance slit.

The spectroscopic quality of the spectra recorded with this optical system can be demonstrated by the example of the determination of six polycyclic aromatic hydrocarbons in the concentration range $0.2–2\,\mathrm{mg\,l^{-1}}$ in a solvent extract of a water sample (Figure 6.4).

The stand-alone design of the detector used in the Perkin-Elmer LC-480 instrument includes all the functions chromatographers need: four-channel detection, analogue or digital on-line output of chromatograms, arithmetic combination of two channels each, e.g. for peak purity confirmation or enhanced selectivity. In addition to the ability of the system to run wavelength programs, the built-in gradient compensation software is an important feature; a baseline drift caused by a solvent gradient can be stored internally for all four channels. As samples are run, the baseline will be corrected on-line and thus can be fed directly into a data-handling device.

The Perkin-Elmer model LC-235 designed for routine applications combines the advantages of a variable UV detector with those of diode array technology. Conversely, the LC 480 AutoScan diode array detector optimizes the sensitivity of self-scanning diode arrays to meet HPLC requirements and facilitate LC–UV coupling by achieving high spectral resolution.

Electrochemical detectors

These are available from several suppliers (Table 6.3). ESA supply the model PS 100A coulochem multi-electrode electrochemical detector. Organics, anions and cations can be detected by electrochemical means.

Sample pretreatment and auto sample injectors

The Gilson Aspec automatic sample preparation system is a fully automated system for solid-phase extraction on disposable columns and on-line HPLC analysis. The Aspec system offers total automation and total control of the entire sample preparation process including clean-up and concentration. In addition, Aspec can automatically inject prepared samples into on-line HPLC systems.

Aspec is designed to receive up to 108 samples. The system is compatible with most standard disposable extraction columns. Analytichem Bond-Elut, Baker SPE, Supelco Supelclean, Alltech Extract Clean, etc. There is a choice of more than 20 different stationary phases.

The Gilson Asted automated sequence trace diazylate enricher (ASTED)

This implements a combination of dialysis and trace enrichment to separate low molecular weight analytes from complex sample matrices. Asted is a novel pre-treatment procedure for biological and industrial samples before HPLC analysis. It removes all interfering macromolecules using a dialysis membrane. The membrane also serves as an efficient filter for microscopic and macroscopic contaminants within the sample. This increases the life of the analytical column and eliminates the need for a guard column. Asted operates on-line with the HPLC system, performing chemical and physical treatments of samples in coordination with the data processor.

The system has been used in amino acid analysis and the analysis of biological samples.

Gilson 231/401 HPLC auto sample injector

The Gilson 231/401 completely automates sampling and injection procedures, increasing throughout while improving precision and minimizing costs. This advanced, robotic unit performs any combination of five operations – dispensing, diluting, mixing, aspirating and injecting – using standard or custom programs to process original samples, diluents, internal and external standards and reagents. The 231/401's components consist of a 231 sample injector, a 401B dilutor and a separate keypad that controls both. This modular structure is adaptable in terms of space and system configuration and allows the 401B to be used separately as a precision dilutor for other applications.

The system holds 80 or 120 sample racks of open or capped samples. Sample volumes between 0.5 and 5 ml can be taken. In addition to non-reactive preparations such as sample dilutions the 231/401 performs reactive sample preparations such as pre-column derivatization. The 231/401 automates pre-column *o*-phthaldialdehyde derivatization of amino acids, with a short analysis time (12 minutes for the separation of 20 amino acids with two internal standards) and high reproducibility.

Gilson 232/401 automatic sample processor and injector

This is a more recent instrument than the 231/401 discussed above. It features automatic sample handling and a built-in column. Its high-capacity sample processor, sophisticated switching valve and powerful injection valve unite to form the most versatile sampler available. Laboratories that routinely analyse large numbers of complex samples can prepare and inject hundreds of samples in a single run. Designed for reliable 24 hour a day operation, the automatic sample processor and injector reduces manual sample manipulation for increased reproducibility and efficiency. Thermostatic racks connected to the circulating water bath accurately maintain sample temperatures indefinitely.

HPLC radioactivity monitor

The Nuclear Enterprise Isoflo system is a microprocessor-based high-efficiency radioactivity monitor designed specifically for use with HPLC. When compared with traditional methods, peak resolution is greatly increased and both time taken and materials used for the measurement are drastically reduced.

Isoflow is similar in many respects to a conventional liquid scintillation counter and the principles of counting are essentially the same for both instruments. In flow monitoring efficiency is maximized, background is minimized and analysis time is optimized for resolution and sensitivity. The chromatographer is able to choose the most appropriate flow cell in order to achieve optimum performance.

Effluent from the column is mixed with a suitable liquid scintillator before being passed through the transparent flow cell for measurement. Alternatively, the effluent may be passed directly through a cell packed with solid scintillator granules compatible with the most commonly used HPLC solvents, thus enabling its recovery for further processing.

6.4.2 Applications

A good example of the application of HPLC in the water industry is the automatic phenol monitoring equipment set up by the North West Water Authority at their Huntington Water treatment plant on the banks of the River Dee. This system incorporates a Varian 5000 series liquid chromatographic system together with a Varian Vista 402 chromatography data system.

The method is based on a trace enrichment technique using reversed-phase chromatography. River water is pumped through a pre-concentration column which retains any phenol and then this column is back-flushed to wash any retained compounds onto the analytical column. Finally, the column eluent is monitored using the Varian UV-100 detector.

Data from the LC system goes to the Varian Vista 402 chromatography data system, where all methodologies as well as raw chromatographic data are stored via the two 5¼ inch floppy disk drives.

Krull *et al.* (1986) have described a method for the determination of inorganic and organic mercury compounds in water by HPLC – inductively coupled plasma emission spectrometry with cold vapour generation. The replacement of the conventional polypropylene spray chamber of the inductively coupled plasma by an all-glass chamber eliminates the severe memory effect obtained when polypropylene is used and enables a detection limit of $32\,\mu g\,l^{-1}$ of mercury to be obtained for mercuric chloride, methyl mercury chloride, ethyl mercury chloride and dimethyl mercury.

6.5 Supercritical fluid chromatography

Until recently the chromatographer has had to rely on either gas chromatographic or HPLC for separations, enduring the limitations of both. Lee Scientific has created a new dimension in chromatography, one which utilizes the unusual properties of supercritical fluids. With the new technology of capillary supercritical fluid chromatography (SFC) the chromatographer benefits from the best of both worlds – the solubility behaviour of liquids and the diffusion and viscosity properties of gases. Consequently, capillary SFC offers unprecedented versatility in obtaining high-resolution separations of difficult compounds.

Beyond its critical point, a substance can no longer be condensed to a liquid, no matter how great the pressure. As pressure increases, however, the fluid density approaches that of a liquid. Because solubility is closely related to density, the solvating strength of the fluid assumes liquid-like characteristics. Its diffusivity and viscosity, however, remain. SFC can use the widest range of detectors available to any chromatographic technique. As a result, capillary SPF has already demonstrated a great potential in application to water, environmental and other areas of analysis.

Suppliers of SFC instruments are reviewed in the Appendix.

SFC is now one of the fastest growing analytical techniques. The first paper on the technique was by Klesper *et al.* (1962), but supercritical fluid chromatography did not catch the analyst's attention until Novotny *et al.* (1981) published the first paper on capillary supercritical fluid chromatography.

SFC finds its applications in compounds that are either difficult or impossible to analyse by liquid chromatography or gas chromatography. SFC is ideal for analysing either thermally labile or non-volatile non-chromatophoric compounds. The technique will be of interest to water chemists as a means of identifying and determining the non-volatile components of water.

Most supercritical fluid chromatographs use carbon dioxide as the supercritical eluent, as it has a convenient critical point of 31.3°C and 72.5 atmospheres. Nitrous oxide, ammonia and *n*-pentane have also been used. This allows easy control of density between $0.2\,g\,ml^{-1}$ and $0.8\,g\,ml^{-1}$ and the utilization of almost any detector from liquid chromatography or gas chromatography.

Wall (1988) has discussed recent developments including timed split injection, extraction and detection systems in SFC.

Time-split injection

Capillary supercritical fluid chromatography utilizes narrow 50 μm or 100 μm i.d. columns of between 3 and 20 m in length. The internal volume of a 3 m × 50 μm i.d. column is only 5.8 μl. Supercritical fluid chromatography operates at pressures from $1500\,lb\,in^{-2}$ to beyond $6000\,lb\,in^{-2}$, this means that GC injection systems cannot be used. HPLC injection systems are suitable for those pressure ranges, but even using small internal loop injectors the volume introduced to the column is very large compared to the column's internal volume. To allow injections of about 10–50 μl to be introduced to a capillary column, an internal loop LC injector (Valco Inst. Switzerland) has been used with a splitter (Figure 6.5(a)) which was placed after the valve to ensure that a smaller volume was introduced onto the column. This method works well for compounds which are easily soluble in carbon dioxide at low pressures.

However, when compounds with high molecular weights are introduced into the system they are often insufficiently soluble to remain in solution in the depressurization area of the split restrictor. The compounds then reprecipitate in the restrictor and cause a decrease in the internal diameter of the restrictor. Hence this reduces the split ratio and causes more compound to be introduced into the column on the next injection, which means that replicate injections show poor reproducibility.

Good reproducibility has been reported for capillary supercritical fluid chromatography using a direct injection method without a split restrictor. This method (Figure 6.5(b)) utilized a rapidly rotating internal loop injector (Valco Inst. Switzerland) which remains in-line with the column for only a short period of time. This then gives a reproducible method of injecting a small fraction of the loop onto the column. For this method to be reproducible the valve must be able to switch very rapidly to put a small slug of sample into the column. To attain this a method called timed-split injection was developed (Lee Scientific). For timed split to operate it is essential that helium is used to switch the valve, air or nitrogen cannot provide sharp enough switching pulses. The injection valve itself must have its internal dead volumes minimized. Dead volumes prior to the valve allow some of the sample to collect prior to the loop, effectively allowing a double slug of sample to be injected which appears at the detector as a very wide solvent peak.

Detection systems

Supercritical fluid chromatography uses detectors from both liquid chromatography and gas chromatography. A summary of detection systems used in supercritical fluid chromatography has been documented (Later *et al.* 1987).

One of the most commonly used detection systems in a gas chromatography laboratory is the electron capture detector. The first paper (Kennedy and Wall 1988) to be published demonstrating the use of an electron capture detector with supercritical fluid chromatograph showed that with supercritical fluid chromatography sensitivity to about 50 pg minimum detection limit on column was obtainable.

A paper has been published showing the use of the photoionization detector (Sim *et al.* 1988). Polyaromatic hydrocarbons are very sensitive using the photoionization detector and the levels detected did not break any new ground in terms of sensitivity. It did inspire HNS Systems (Newtown MA, USA), who market a photoionization detector, to try the detector with a capillary system, interfaced to a Lee Scientific 602 supercritical fluid chromatograph (Lee Scientific, Salt Lake City, Utah, USA).

The photoionization detector is to a certain extent specific in that only compounds that can be ionized by a UV lamp will give a response. The solvents used were dichloromethane and acetonitrile, both of which should have little response in the photoionization detector. However, a clear sharp solvent peak was observed.

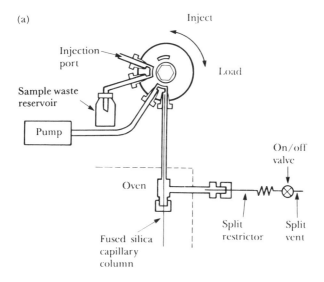

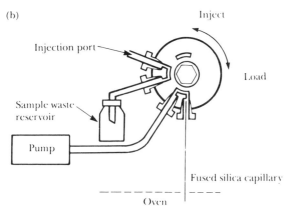

Figure 6.5 *Sample injectors: (a) split valve inspector; (b) timed split and direct valve injector*

The amount detected by this system (0.3 pg on column) was below the level which could have been determined using a flame ionization detector. Initial indications show that the photoionization detector may be a very useful detector for people who wish to get to lower levels on the supercritical fluid chromatograph and cannot concentrate their sample.

Sulphur chemiluminescence detector

The flame photometric detector commonly used in gas chromatography for sulphur specific detection has found little application in supercritical fluid chromatography. Flame photometric detection is not used in SFC because the initial results obtained with SFC using a flame photometric detector showed the response for sulphur-containing species to be very poor. Carbon dioxide has a coincident emission line which cannot be resolved from the main sulphur line, making flame photometric detection almost useless with SFC and carbon dioxide. Other sulphur detectors do exist, such as the sulphur chemiluminescence detector (CD) (Sievers Research Inc. Colorado, USA). The link to supercritical fluid chromatography has been investigated. Good sensitivities and chromatograms have been shown for standards and real samples. This detector shows no response to carbon dioxide and gives low picogram sensitivities for a wide range of sulphur compounds.

The newest developments in supercritical fluid chromatography instrumentation are the Lee Scientific 602 SFC and 622 SFC/GC. These incorporate the latest advanced technology, the latter being a dual-purpose SFC gas chromatographic instrument. They feature a pulseless high-capacity pump, a high-temperature oven for SFC and gas chromatography, compatibility with packed and capillary columns, high-sensitivity detectors (flame ionization, UV, FTIR and MS) and newly developed software capable of creating an infinite variety of simultaneous temperature and density or pressure programmes.

In Figure 6.6 are shown some SFC chromatograms obtained using the Lee Scientific 501 SFC instrument.

6.6 Gas chromatography mass spectrometry

The time has long since passed when one could rely on gas chromatographic or liquid chromatographic data alone to identify unknown compounds in water or other environmental samples. The sheer number of compounds present in such materials would invalidate the use of these techniques, and even in the case of simple mixtures the time required for identification would be too great to provide essential information in the case, for example, of accidental spillage of an organic substance into a water course or inlet to a water treatment plant where information is required very rapidly.

The practice nowadays is to link a mass spectrometer or ion trap to the outlet of the gas chromatograph or high-performance liquid chromatograph so that a mass spectrum is obtained for each chromatographic peak as it emerges from the separation column. If the peak contains a single substance then computerized library-searching facilities attached to the mass spectrometer will rapidly identify the substance. If the emerging peak contains several substances, then the mass spectrum will indicate this and in many cases will provide information on the substances present.

The use of gas chromatography/mass spectrometry grew rapidly during the early 1970s as discussed by Shackleford and McGuire (1986).

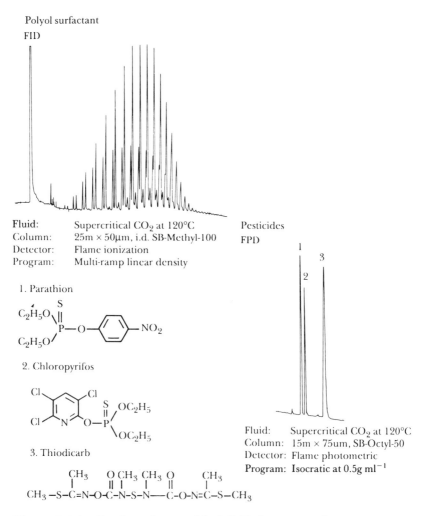

Polyol surfactant

FID

Fluid: Supercritical CO_2 at 120°C
Column: 25m × 50μm, i.d. SB-Methyl-100
Detector: Flame ionization
Program: Multi-ramp linear density

1. Parathion

2. Chloropyrifos

3. Thiodicarb

Pesticides

FPD

Fluid: Supercritical CO_2 at 120°C
Column: 15m × 75um, SB-Octyl-50
Detector: Flame photometric
Program: Isocratic at 0.5g ml^{-1}

Figure 6.6 *Applications of supercritical fluid chromatography*

The first large-scale application of gas chromatography/mass spectrometry to analysis of environmental pollutants occurred in 1977 when the effluent guidelines division of EPA, under court order, began collecting data and writing regulations to limit the discharge of pollutants into surface waters. Tellaid (1986) and others (Shackleford and McGuire 1986; Lichtenberg *et al.* 1986; Federal Register 1979, 1986, 1984b; Colby 1986; Fisk *et al.* 1986; Friedman 1986) give a history of the selection of the EPA priority pollutants, the selection of gas chromatography/mass spectrometry as the technique of choice for their analysis and the problems faced in moving a research technique into production.

The EPA (Federal Register 1984b) published gas chromatography mass spectrometry methods for the examination of solid wastes (Federal Register 1984a).

In addition to the processing work of the EPA many individual laboratories throughout the world concerned with the analysis of water and many other environmental materials (fish, effluents, sediments, etc.) have set up equipment for gas chromatography mass spectrometry, and what was once considered to be an expensive instrument purchased perhaps to solve a single vitally important problem has become a general workhorse instrument. No water laboratory which aims to be able to solve the kinds of problems thrust upon it can afford to be without this technique. Instead, the problem is one of choosing the most appropriate instrumentation for their needs, and this is discussed further below.

6.6.1 Instrumentation

Finnigan MAT are the main suppliers of this equipment. Equipment available from this and other companies is listed in the Appendix.

SSQ 70 series single-stage quadrupole-mass spectrometer

This offers premium single-stage performance, with the option of being upgraded to a triple-stage quadrupole system (i.e. the TSQ 70). The SSQ 70 features a network of distributed microprocessors with more than 1.5 megabytes of memory linked to a powerful DEC 11/73 processor with 2.0 megabytes of memory for data-processing operations. Instrument control tasks can be displayed in up to eight windows on a colour display terminal. The hyperbolic quadrupole analyser gives the SSQ 70 a mass range of up to 4000 μm; system performance is specified to 200 m/z. The cradle vacuum system with three large inlet points at the ion source accommodates a variety of sample inlets such as capillary gas chromatography, thermospray, liquid chromatography, mass spectrometry, supercritical fluid chromatography and solids probe. Standard features of the instrument also include high-performance EI/CI (electron impact/chemical ionization) ion-source with exchangeable ion volumes, a PPI NICI with high-voltage conversion dynode multiplier for positive and negative ion detection and fast ion bombardment. The Varian 3400 GC gas chromatograph incorporates a high-performance capillary column with multilinear temperature programming in up to 8 sequences, a data-control and recording system for temperatures in the gas chromatograph oven and for interface temperatures, and also for controlling and recording value timing, a data system control of optional gas chromatography accessories and a split/splitless capillary injector. The Micro VIP computer data system comprises a DEC 11/73 processor with video colour display, dot matrix printer and a data system for control of instrument control parameters and user-initiated diagnostics.

Mass spectrometry–mass spectrometry

In high-performance mass spectrometry–mass spectrometry (as opposed to gas chromatography–mass spectrometry) the separator as well as the analysis is performed by the mass spectrometer. One advantage of this technique over combined chromatography–mass spectrometry is that separation is a spatial process rather than being dependent on time. This can lead to improved analysis times and/or greater specificity. Mass spectrometry–mass spectrometry also opens up other areas such as the study of complete structures. This technique has been discussed in detail by Warburton and Millard (1980).

H-SQ 30 hybrid mass spectrometer–mass spectrometer

This instrument combines a reverse-geometry (BE) magnetic section instrument with a quadrupole (QQ) analyser. This hybrid combination provides mass spectrometry–mass spectrometry operation with a high-resolution first stage (BE) and a unit resolution second stage (QQ). The four available collision regions allow experiments of low (2–100 eV) and high (3 keV) collision energy, as well as consecutive CID experiments using two separate collision regions. The H-SQ 30 is an ideal instrument for structural elucidation studies and ion physics.

MAT-90 high-mass–high-resolution mass spectrometer

This is a very high-performance instrument in which instrument control resides in a multiprocessor system manager leaving only the analytically important parameters to be defined by the operator. It utilizes a completely new concept of ion optics for double focusing and this gives the instrument unmatched performance.

The performance of a magnetic sector mass spectrometer depends totally on the ability to focus ions from source to detector. To produce ideal focusing a very wide range of factors must be taken into account. Modern computer simulation techniques have now been extensively applied in this instrument and have resulted in an ion optical design closer to the ideal than ever before. This configuration provides for complete image error correction in all planes.

System resolution in excess of 50 000 is achieved and excellent performance is obtained at high masses. The instrument features a novel ion source which can be exchanged in a few seconds via vacuum lock. Optimized EI and CI systems are supplied.

Optional ionization volumes are available for fast ion bombardment and alternating CC/EI. The MAT-90 analyser has reverse Nier–Johnson geometry allowing metastable studies to be carried out using both first-order and second-order field free regions in the standard system. To extend the application of metastable techniques, the optional collision cell in the first field-free region can be used. A full range of accessories is available, including direct-probe, fast direct-probe, thermospray on-line mass spectrometry, automatic direct evaporation, fast ion bombardment, direct chemical ionization and continuous-flow fast ion bombardment.

FD/FI device

The standard MAT 90 ion source is used for optimized FD/FI mode by means of the newly designed FD/FI probe. Conversion from electron impact (EI), chemical ionization (CI) or fast ion bombardment (FAB) to FD/FI operation does not require the exchange of the ion source. The FD/FI probe accommodates both the field emitter and the extraction electrodes, mounted at the probe tip. Both are introduced as a unit into the ion source through the ionization volume exchange lock without breaking vacuum. The fast and simple changeover illustrates the versatility of the MAT 90 with no compromise on the performance.

All aspects of system control, data acquisition and processing are carried out in the integrated multiprocessor system. The primary processor is a DEC POP 11/73 with 2 megabytes (optional up to 4 megabytes) of main memory with cache and disk cache memory. The user interface is via a high-resolution colour terminal. Standard features also include computer-controlled variable entrance and exit slits, electron multiplier with $\pm 20\,kV$ diode, direct coupled capillary gas chromatography–mass spectrometry interface with precise temperature control up to 400°C data system software including library search, quantification and data handling.

Series 700 ion-trap detector

The ion-trap detector detects any compound that can be chromatographed; it is a universal detector that can replace several conventional gas chromatography detectors such as the type used in the Varian model 3400 gas chromatogram included in the Finnigan MAT SSA-70 and TSQ-70 instruments. Electron capture, flame ionization, element specific (etc.) detectors used in the latter instruments are not universal in this sense and will not respond to all types of organic compounds, i.e. some compounds will be missed. The ion-trap detector obviates this difficulty by responding to all types of organic compounds. In the ion-trap technique one does not have to rely on retention data for identification. The mass spectrum tells you the identity with certainty. Unidentified gas chromatographic peaks are a thing of the past. Complete analysis and identification is done in one run with one detector. This makes the ion-trap detector a very attractive proposition to the water chemist. Various aspects of ion-trap detectors have been discussed by workers at Finnigan MAT and elsewhere. The following references give the appropriate Finnigan MAT IDT numbers: Kelly (10, 21); Campbell (15); Stafford (16, 20, 24); Rordorf (13); Syka (19); Yost *et al.* (22); Camp (23); Richards and Bradford (25); Bishop (28, 36, 42); Campbell and Evans (29); Olsen (35); Allison (41); Todd *et al.* (46); Eichelberg and Buad (47); Eichelberg and Slivon (48); Genin (53); Leheir (51); Richards *et al.* (56). See also Finnigan MAT ADS 7–10, 12–14, 24, 27, 29.

During development of the ion-trap detector, it was found that the low voltage previously used for storage encouraged the production and storage of the H_2O^+ and H_3O^+ ions which occasionally led to an increase of the $M + 1$ molecular ions. This problem had been eliminated by adjusting the storage voltage such that the H_2O^+ and H_3O^+ ions are no longer stored.

This scanning method produces standard electron impact spectra which can be rapidly searched through the standard NBS library (42 222 spectra). In each case the number of ions stored in the trap would be the optimum required to produce a conventional electron impact spectrum. In order that the procedure will not affect quantitative results, ion intensities are stored after application of an adjustment factor which is always related to the true size of the peak as measured by the original fast scan. Scaling is controlled by the computer and the net result is a system with a dynamic range between 10^4 and 10^5. The efficiency of the procedure has been evaluated by measuring the signals obtained from injections of a difficult compound in quantities ranging from 10 pg to 10 ng. Each result has been measured three times and a log/log plot of signals against concentration shown to be a straight line over the entire range with a correlation coefficient >99%.

The ion-trap detector may be operated both as a universal detector (when full scans are stored) or, with the application of multiple ion monitoring, as a specific detector. Because approximately 50% of the ions formed in the trap are analysed, the sensitivity of the instrument in full-scan mode can be much higher than conventional mass spectrometers, in which only 0.1–0.2% of the ions formed may be detected. Thus the instrument can be used to detect 2–5 pg (in full scan) of compounds eluting from the column; a performance which compares extremely favourably with those of the most sensitive specific detectors (e.g. the electron capture detector) and easily outstrips that of the flame ionization detector. As already indicated, this sensitivity is not achieved at the expense of dynamic range; as the instrument can produce linear calibration graphs for quantities within the range 5–10 pg to 1000 ng on column. This again compares favourably with the performance of the flame ionization detector.

When operated as a specific detector the ion-trap detector is more sensitive still but not to the extent that would be expected from the performance of other mass spectrometers operated in this mode; in view of the large number of ions monitored in full scan mode there is little more sensitivity to be gained by spending a little extra time scanning a narrow mass range, and the detection limit in this mode is in the region of 1–2 pg.

The power of the system to overcome the problems associated with co-eluting compounds is demonstrated in conjunction with the use of deuterated (or ^{13}C-labelled compounds) as internal standards. Such techniques could not be used in conventional gas chromatography as the deuterated compounds often co-elute, making quantification difficult if not impossible. With the ion-trap detector, however, it is easily possible to differentiate between the ions arising from the different compounds and the intensities of these ions could then be used for quantification of the compounds involved. The application of such techniques can be shown by the quantitation of anthracene. Chromatograms of ions characteristic of anthracene and its deuterated analogue (*m/e* 178 and 188 respectively) indicated that the compounds did not in fact elute simultaneously. The retention time of the labelled compound was fractionally less than that of the unlabelled material.

In Figure 6.7(a) is shown a partial chromatogram of a complex mixture of chlorinated biphenyls extracted from a water sample.

The signals from masses 292 and 326 characteristic of tetra- and pentachlorobiphenyl are shown in Figure 6.7(b,c). The specific detection mode of the ion-trap detector can be used to improve detection limits. This

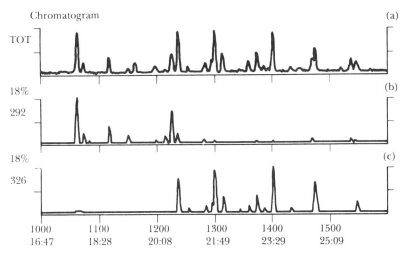

Figure 6.7 *Specific detection. Trace (a) shows a partial chromatogram for Arochlor 1254. Trace (b) shows the mass chromatogram for 292 characteristic of tetrachlorobiphenyls in the mixture. Trace (c) shows the mass chromatogram for 326 characteristic of pentachlorobiphenyls*

detector can monitor specific masses that are characteristic of compounds of interest. The detector records the signal for only those masses and ignores all others. Interference from other compounds is virtually eliminated with the Finnigan MAT 700 detector – up to 16 different groups of masses can be monitored or a mass range of up to 40 masses can be handled. With this flexibility it is possible to monitor only the masses of interest and to improve detection limits.

Incos-50 quadrupole mass spectrometer

The Incos-50 is a relatively low-cost benchtop instrument as opposed to the research grade instruments discussed earlier. The gas chromatograph mass spectrometer transfer lines allow it to be used with either the Hewlett Packard 5890 or the Varian 3400 gas chromatographs. The Incos 50 provides data system control of the gas chromatograph and accessories such as autosampler or liquid sample concentration. It can be used with capillary, wide-bore or packed columns. It performs electron ionization or chemical ionization with positive or negative detection. It also accepts desorption or other solids controls.

Finnigan MAT Chem Master Workstation

The Chem Master Workstation is a gas chromatography and gas chromatography mass spectrometry data-processing system that speeds the flow of data through the laboratory and provides essential quality-assurance and quality-control review. It is a PC-based integrated hardware/software system that converts gas chromatographic and gas chromatography mass spectrometric data into reliable analytical reports.

Model 1020 routine gas chromatograph mass spectrometer

This is a cost-effective completely automated system optimized for the routine analysis of complex organic samples. It is specifically designed to meet the needs of an analytical laboratory requiring a gas chromatograph mass spectrometer with the following characteristics:

- high sample throughput
- low initial investment
- low operating costs
- ease of automation
- complete software package
- serviceability
- field-proven hardware

The model 1020 software package includes interactive programs specifically designed for complex mixture analysis and advanced automated programs for routine analysis. All system functions are computer controlled with minimal knowledge of mass spectrometry.

All gas chromatograph parameters, including temperature program rates and hold times, are controlled by the microprocessor and set through the CRT keyboard. Up to five sets of parameters can be stored on the computer disk for instant recall.

The mass spectrometer, when combined with a computer data system, precisely identifies and quantifies each sample component as it elutes from the gas chromatograph. The model 1020 uses an electron ionization source to produce mass fragments, and a quadrupole mass filter, one of the most accurate and cost-effective devices for separating mass fragments.

For most applications, excellent spectra are produced with only a few nanograms of sample. When performing single ion monitoring picogram and femtogram levels of many compounds can be precisely quantified. In addition to single-ion monitoring, the system's powerful software permits multiple ion monitoring of up to 25 mass ranges. This improves sensitivity and reliability of compound identifications by allowing a combination of multiple-ion mass ranges as well as single-ion monitoring to be performed in a single analysis. Mass stability of better than $\pm 0.1\,\mu g$ per day ensures accurate mass assignment.

OWA -20/30B organics in water gas chromatograph mass spectrometer

This system combines hardware and software features not found in any other low-cost gas chromatography mass spectrometry system. The highly reliable 3000 series electron ionization source and quadrupole analyser are used to provide superior mass spectrometer performance. The software is designed with the necessary automation to perform complete quantitative analysis of any target compounds. All routine system operating parameters are adjustable through the computer's graphics display terminal. The priority interrupt foreground/background operation system allows all data-processing functions to be performed at any time with no limiting effects on data acquisition. Sophisticated data-processing programs are readily accessible through a simple commercial structure. The simplicity of the entire system allows complete analysis with minimal operator training. Standard features of this instrument include fully automated gas chromatography mass spectrometry, automated compound analysis and quantification, software, $4-800\,\mu$ electron impact quadrupole mass spectrometer, high-capacity turbidmolecular pump vacuum system, liquid sample concentrators for volatile organics in water analysis, a sigma series programmable gas chromatograph, grob-type split–splitless capillary column injector system, packed column injector with glass jet separator, Nova 4C/53K word, 16-bit minicomputer, graphics display terminal, 10-megabyte disk drive, a

Table 6.4. *Applications of gas chromatography mass spectrometry to water analysis*

Title	Types of compounds	Reference
The content decree pollutants and their analysis by GC/MS	–	Tellaid (1986)
Analysis of water extractable priority pollutants by GC/MS	Methylene dichloride, extractables, e.g. phenols, substd. phenols hydrocarbons, insecticides, phthalates, PAH, PCB	Shackleford and McGuire (1986)
GC/MS method for the determination of priority pollutants in water by the purge and trap method	Aromatics, aliphatics, halohydrocarbons, vinyl halides	Lichtenberg *et al.* (1986)
GC/MS determination of pollutants in the Tojiang River in China	182 organic compounds detected	Quin *et al.* (1986)
GC/MS analysis of organic compounds in Beijing rainwater using the ITD	100 organic compounds detected	Rugin *et al.* (Finnigan-MAT App. Rep. 215)
Isotope dilution GC/MS methods for determining priority pollutants in water	Methylene chloride extractables and purgeables as listed above	Colby (1986)
The use of quadrupole and ion-trap mass spectrometers for the identification of air and water pollutants	27 halohydrocarbons and aromatics (purge and trap)	Evans *et al.* (1984)

Analysis of volatile organic compounds in industrial wastes by the Finnigan OWA GC/MS	27 halohydrocarbons and aromatics (purge and trap)	Schnute (Finnigan Corp. App. Rep. AR8018)
GC/MS method for analysis of pollutants in hazardous wastes	Volatile (purgeable) semi-volatile (solvent extractable insecticide/PCB (residue) fractions in soil, water and sediments (126 compounds mentioned)	Fisk *et al.* (1986)
GC/MS methods for the analysis of solid waste	Wide variety of organics in sludge or waste	Friedman (1986)
Analysis of polychlorinated biphenyls by gas chromatography – ion-trap detector	–	Shek (Finnigan-MAT IDT (4))
Analysis of chlorinated insecticides using the ion-trap detector	–	Wellby (Finnigan-MAT IDT (6))
Use of pullutant and biogenic markers as source discriminants of organic inputs to estuarine sediments	–	Steadman and Mantura (Finnigan-MAT IDT (8))
The use of quadrupole and ion-trap mass spectrometers for the identification of air and water pollutants	–	Evans and Smith (Finnigan-MAT IDT (11))
Comparison of quantitative results for analysis of 2, 3, 7, 8 TCDD in fish by 4500 quadrupole and 705 ion-trap detector	–	Rordorf (Finnigan-MAT IDT (14))
Analysis of contaminated water using the Finnigan-MAT ion-trap detector	–	Ster (Finnigan-MAT IDT (17))

Table 6.4. (*Continued*)

Title	Types of compounds	Reference
Mass spectra of fatty acid derivatives of isopropylidenes of novel glycerylethers of cod muscle and of phenolic acetates obtained with the Finnigan-MAT ion trap detector	—	Ratnayake (Finnigan-MAT IDT (33))
Determination of polychlorinated biphenyl substitute Ugilec (tetrachlorobenzyl toluenes) in fish	—	Furst et al. (Finnigan-MAT IDT (43))
Trace amounts of polychlorinated dibenzo-p-dioxins with the ion-trap detector	—	Finnigan-MAT AD 511
Gas chromatographic analysis of phthalate esters with the ion trap	—	Finnigan-MAT AD 514
Determination of polychlorinated biphenyls in industrial waste streams with IDT	—	Finnigan-MAT AD 522
Trace-level analysis of tetra-chloro-dibenzo-p-dioxins with IDT	—	Finnigan-MAT ADS 26

printer/plotter, an NBS 31000 spectra library, a full scan or multiple ion detector and a 9-track tape drive. Options include chemical ionization ion source, direct inlet vacuum lock, programmable solids probe, direct exposure probe, various GC detectors, autosampler, subambient GC operation and a 32-megabyte disk drive.

6.6.2 Applications

Some of the many applications of the technique are summarized in Table 6.4.

As an example of the application of gas chromatography mass spectrometry, Figure 6.8 shows a reconstructed ion chromatograph obtained for an industrial waste sample. The Finnigan MAT 1020 instrument was used in this work. Of the 27 compounds searched for, 15 were found. These data were automatically quantified. This portion of the report contains the date and time at which the run was made, the sample description, who submitted the sample and the analyst, followed by the names of the compounds. If no match for a library entry was found, the component was listed as 'not found'. Also shown is the method of quantification and the area of the peak (height could also have been chosen).

The large peak at scan #502 (Figure 6.8) does not interfere with the ability of the software to quantify the sample. Although the compound eluting at

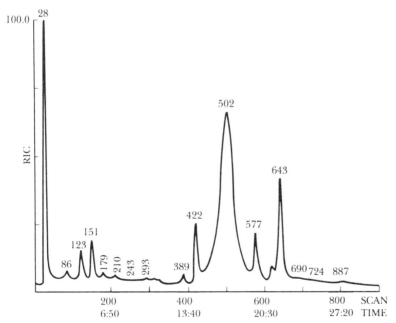

Figure 6.8 *Reconstructed ion chromatograph of industrial waste sample*

scan #502 was not one of the target compounds in the library being reverse-searched, it was possible to identify it by forward-searching the NBS library present on the system. The greatest similarity was in the comparison of the unknown with the spectrum of benzaldehyde.

6.7 High-performance liquid chromatography mass spectrometry

One of the limitations of gas chromatography and consequently of gas chromatography mass spectrometry is that of all the organic material present in natural water samples, only a small proportion, say as low as 20%, is sufficiently volatile to be separated on gas chromatographic columns operating at even the maximum of their temperature range. Of the 275 compounds for the Appendix III list of the US Environmental Protection Agency, 150 are not amenable to gas chromatographic separation.

As a consequence of this there has, in recent years, been a growing interest in applying high-performance liquid chromatography which is not subject to this temperature limitation, to the determination of the non-volatile fractions of water.

6.7.1 Instrumentation

Hewlett Packard supply the HP 5988A and HP 5987A mass-selective detectors for use with liquid chromatographs.

To date this equipment has been used extensively for identifying and determining non-volatile compounds such as diuretics and some stimulants in urine samples taken at the Olympic Games and the technique is now being introduced into the water laboratory. Particle beam technology has produced further improvements in liquid chromatography mass spectrometry. The particle beam liquid chromatograph mass spectrometer uses the same switchable electron impact chemical ionization source and the same software and data systems that are used for a gas chromatography mass spectrometry system. Adding a gas chromatograph creates a versatile particle-beam liquid chromatography/gas chromatography/mass spectrometry system that can be switched from liquid chromatography mass spectrometry to gas chromatography mass spectrometry in an instant.

Based on a new technology, particle beam enhanced liquid chromatography/mass spectrometry expands a chemist's ability to analyse a vast variety of substances. Electron impact spectra from the system are reproducible and can be searched against standard or custom libraries for positive compound identification. Chemical ionization spectra can also be produced. Simplicity is a key feature. A simple adjustment to the particle beam interface is all it takes.

The particle beam system is a simple transport device, very similar to a two-stage jet separator. The solvent vapour is pumped away, while the analyte particles are concentrated in a beam and allowed to enter the mass spectrometric source. Here they are vapourized and ionized by electron impact.

The different ways a particle beam liquid chromatograph mass spectrometer can be configured reflect the versatility of the system in accommodating both the application and the availability of existing instrumentation. The system consists of these elements:

- particle beam interface mounted on the Hewlett Packard 5988A or 5987A mass spectrometer
- liquid chromatograph (either the integrated Hewlett Packard 1090 or modular Hewlett Packard 1050)
- data system (either HP 59970C chem station for single instrument operation or the Hewlett Packard 1000 RTE A-series for multi-instrument, multi-tasking, multi-user operation)

This technique is complementary to the thermospray technique. Relative advances of the particles beam technique over thermospray include library searchable electron impact spectra, improved reproducibility, easier use and increased predictability over a broad range of compounds. But since a particle beam requires some sample volatility, very large and polar compounds such as proteins may not provide satisfactory results using particle beam liquid chromatography mass spectrometry. Additionally, certain classes of compounds such as preformed ions, azo dyes and complex sugars may not yield satisfactory electron impact spectra, but can be run on thermospray. In other words, both liquid chromatography mass spectrometry techniques complement each other's limitations and the analyst may want to add both to address a broader range of samples.

6.7.2 Applications

Liquid chromatography mass spectrometry has been used to determine triazine pesticides in land-fill sites (Hewlett Packard 1988), alkylbenzene sulphonates, polyethylene glycols and nonylphenylethoxylates in potable water and dioctadecylmethyl ammonium in sewage effluent (Rivera *et al.* 1985; Righton and Watts 1986).

6.8 Infrared and Raman spectroscopy

These techniques have limited applications in the water laboratory. Thus infrared spectroscopy has been used for determining total hydrocarbons in carbon tetrachloride extracts of water samples. A more recent development is Fourier transform infrared analysis.

Fourier transform infrared spectroscopy

Fourier transform infrared spectroscopy, a versatile and widely used analytical technique, relies on the creation of interference in a beam of light. A source light beam is split into two parts and a continually varying phase difference is introduced into one of the two resultant beams. The two beams are recombined and the interf‹rence signal is measured and recorded, as an interferogram. A Fourier transform of the interferogram provides the spectrum of the detected light. Fourier transform infrared spectroscopy, a seemingly indirect method of spectroscopy, has many practical advantages, as discussed below.

A Fourier transform infrared spectroscopy spectrometer consists of an infrared source, an interference modulator (usually a scanning Michelson interferometer), a sample chamber and an infrared detector. Interference signals measured at the detector are usually amplified and then digitized. A digital computer initially records and then processes the interferogram and also allows the spectral data that results to be manipulated. Permanent records of spectral data are created using a plotter or other peripheral device.

The principal reasons for choosing Fourier transform infrared spectroscopy are first, that these instruments record all wavelengths simultaneously and thus operate with maximum efficiency and second, that Fourier transform infrared spectroscopy spectrometers have a more convenient optical geometry than do dispersive infrared instruments. These two facts lead to the following advantages:

- Fourier transform infrared spectroscopy spectrometers achieve much higher signal-to-noise ratios in comparable scanning times.
- Fourier transform infrared spectroscopy spectometers can cover wide spectral ranges with a single scan in a relatively short scan time, thereby permitting the possibility of kinetic time-resolved measurements.
- Fourier transform infrared spectroscopy provides higher-resolution capabilities without undue sacrifices in energy throughput or signal-to-noise ratios.
- Fourier transform infrared spectrometers encounter none of the stray light problems usually associated with dispersive spectrometers.
- Fourier transform infrared instruments provide a more convenient beam geometry – circular rather than slit shaped – at the sample focus.

Fourier transform Raman spectroscopy

Conventional Raman spectroscopy cannot be applied directly to water samples, although it is occasionally used to provide information on organic solvent extracts of water samples. Fourier transform Raman spectroscopy, on the other hand, can be directly applied to water samples (*Laboratory News* 1988). The technique complements infrared spectroscopy in that some functional groups (e.g. unsaturation) give a much stronger response in the Raman region whilst others (e.g. carbonyl) give a stronger response in the infrared. Several manufacturers (Perkin-Elmer, Digilab, Bruker) are currently developing Fourier transform infrared spectrometers.

6.9 Amino acid analysers

This is an example of a dedicated application of high-performance liquid chromatography. Measurements of amino acids have been performed in lake water (Gardner and Lee 1973) and sea water (Dawson and Pritchard 1978; Bajor and Bohling 1970; Tusck *et al.* 1979). Also, measurements of total protein or amino acids are required in river waters, effluents, milk, blood, wine, etc.

The most popular current techniques for amino acid analysis rely on liquid chromatography and there are two basic analytical methods. The first is based on ion-exchange chromatography with post-column derivatization. The second uses pre-column derivatization followed by reversed-phase HPLC. Derivatization is necessary because amino acids, with very few exceptions, do not absorb in the UV–visible region, nor do they possess natural fluorescence.

Each of the major methods has its own particular advantages and disadvantages. Since the variety of available chemistries can be confusing, the method itself should govern the choice that meets requirements, rather than the equipment or systems offered. The optimal method is best selected by a comprehensive and objective review of all commonly used techniques (Table 6.5).

Over-riding criteria which will influence the selection are resolution, sensitivity and speed. Whilst the very best chromatogram for any given method will inevitably be a compromise, only the fullest evaluation of all the alternatives offered will guarantee a correct selection. The various manufacturers of this equipment are listed in the Appendix.

Certainly, a vast amount of experience has been gained by the widespread use of conventional amino acid analysers. They offer high reliability, accuracy, reproducibility and can separate complex samples. Because conventional analysers can be fully automated, they are widely used in routine analysis. However, the method is limited by the sensitivity which can be achieved using ninhydrin as the derivatizing agent. Sensitivity can be increased by using ortho-phthaldialdehyde (OPA) instead, but where extremely high sensitivity is required, HPLC is the method of choice.

Two other reagents used in HPLC are 9-fluorenyl methoxycarbonyl chloride (FMOC) and phenylisothiocyanate (PITC). Fluorenyl methoxycarbonyl chloride is becoming increasingly popular in protein chemistry

Table 6.5. *Procedure for the determination of amino acids*

Form of chromatography	Derivatization	Reagent		
Ion exchange	Post-column	Ninhydrin	Amino acids are separated in their native form on a sulphonated polystyrene resin using a system of sodium or lithium based buffers. Separation is effected by stepwise, rather than gradient elution, and the chromatography can be further optimized by carefully controlling the temperature of the analytical column. The number and nature of the buffers used depends on the complexity of the sample to be analysed. Hydrolysates of protein and foodstuff samples can be separated on a 3 or 4 buffer sodium system, whereas physiological fluids (which are inherently more complex) require the use of 5 lithium buffers. The post-column ninhydrin reaction proceeds in a reaction coil at elevated temperatures (ca. 130–135°C). Ninhydrin reacts with primary amino acids to give a chromophore with a wavelength of maximum absorption of 570 nm. With secondary amino acids, ninhydrin reacts to form a yellow complex with different absorbance characteristics. For this reason, detection occurs at both 570 and 440 nm.	 Ninhydrin　Amino acid　Ruhemann's purple *Simplified reaction between ninhydrin and primary amino acids* Ninhydrin　Proline　Yellow chromophore *Simplified reaction between ninhydrin and secondary amino acids*	

| Ion exchange | Post-column | o-phthal-dialdehyde (OPA) | The reaction between OPA and primary amino acids takes place in the presence of a thiol such as 2-mercaptoethanol, to form highly fluorescent derivatives. By using OPA as the derivatizing reagent, the sensitivity of a conventional analyser can be increased by a factor of 10.

On the other hand, OPA does not react with secondary amines and is therefore unsuitable if detection of proline or hydroxyproline is essential. OPA also has a low response with cystine and other disulphides.

The OPA reaction proceeds rapidly at room temperature and the post-column fluidics are simplified by replacing the existing reaction module and photometer with a fluorescence detector. Excitation occurs at 330 nm, and emission is measured at a wavelength of 430 nm or above. | $\text{OPA} + NH_2CHRCO_2H + R'SH \rightarrow$ Fluorescent adduct

OPA   Amino acid   Thiol   Fluorescent adduct

Simplified reaction between OPA and primary amino acids |
| Reversed phase | Pre-column | o-phthal-dialdehyde | OPA, unlike ninhydrin (and in common with other precolumn derivatizing agents), forms a different adduct with each amino acid. The resulting derivatives can be separated using reversed-phase chromatography.

OPA-derivatized amino acids are usually separated on an ODS-II solid phase using a mobile phase of sodium phosphate buffer and an acetonitrile gradient.

In contrast to the post-column derivitization, the inherent sensitivity of this method is higher since the column effluent is not diluted. Adequate sensitivity can therefore be achieved by UV monitoring at 330 nm. For higher sensitivity, fluorescence detection is often chosen, and emission is measured at wavelengths above 430 nm. Moreover, baseline quality is superior.

There is no necessity to remove excess OPA prior to sample injection since OPA itself will not interfere with separation or detection. However, since OPA-amino acid derivatives are unstable, complete automation of the precolumn reaction with accurate control of reaction time is essential for reproducible results. | $\text{OPA} + NH_2CHRCO_2H + R'SH \rightarrow$ Fluorescent adduct

OPA   Amino acid   Thiol   Fluorescent adduct |

Table 6.5. (*Continued*)

Form of chromatography	*Derivatization*	*Reagent*	
Reversed phase	Pre-column	Phenyl-isothio-cyanate (PITC)	PITC has been used extensively in the sequencing of peptides and proteins and reacts under alkaline conditions with both primary and secondary amino acids. The methods of sample preparation and derivitization follow a stringent procedure which involves many labour-intensive stages. However, the resulting phenylthiocarbamyl-amino acids (PTC-AA's) are very stable, and the timing of the derivatization step is not as critical as when using OPA. Hardware requirements for the separation of PTC-amino acids are similar to those for OPA amino acid determination. Since the derivatives do not fluoresce the technique is limited to UV detection, which normally takes place at 245 nm.

Despite the problems encountered in sample preparation, the chromatography of protein hydrolysate samples is straightforward and can give good resolution in a short time.

PITC   Amino acid   PTC – amino acid

Simplified reaction between PITC and amino acids |

Reverse phase	Pre-column	9-fluo-renyl methoxy-carbonyl chloride (FMOC)	FMOC is a relatively new reagent which has not only made a very significant impact in peptide synthesis, but has also become used as an effective pre-column derivatizing reagent in amino acid analysis.

FMOC offers sensitivity comparable to OPA when used in conjunction with fluorescence detection, whilst also reacting with both primary and secondary amines. Moreover, the derivatives are stable for more than 30 hours in an acidic environment.

Both FMOC and its hydrolysis products have similar absorption and fluorescence spectra to FMOC-amino acids. Excess FMOC remaining after derivatization reacts with water to form 9-fluorenylmethyl alcohol (FMOC-OH), and if this is not removed prior to sample injection, it elutes as a large, broad peak in the vicinity of proline. Removal of FMOC-OH can be achieved by liquid–liquid extraction (normally into pentane). A more elegant method of preventing interference of FMOC-OH is to react excess FMOC with a very hydrophobic amine (e.g. 1-aminoadamantane) to form a derivative which elutes after the peaks of interest.

FMOC-amino acids can be chromatographed using a C8 column and acetonitrile in sodium acetate buffer as the mobile phase. Fluorescence detection with excitation at 260 nm and emission at 310 nm gives the best results.

FMOC CH_2OCOCl + NH_2CHRCO_2H Amino acid $\rightarrow$ $CH_2OCONHCHRCO_2H$ FMOC – amino acid

Simplified reaction between FMOC and amino acids

research because it reacts with secondary amines and also offers rapid analysis of protein hydrolysates.

One aspect governing the choice of method is the sensitivity required. If only a small amount of sample is available, then for the LKB Alpha Plus and the LKB HPLC instruments, the greatest sensitivity is obtained using the Alpha Plus instrument with ion-exchange separation and post-column derivatization with *o*-phthaldialdehyde (OPA) reagent (108 µg) or the HPLC instrument using reversed-phase chromatography and pre-column derivatization with OPA, or 9-fluorenylmethoxycarbonyl chloride (FMOC) reagents (30–33 ng).

6.9.1 Instrumentation

One leading supplier, LKB, is discussed below. Others are reviewed in the Appendix.

LKB supply two instruments, the LKP 4150 Alpha HPLC, and for analysis requiring higher sensitivity and faster run times the LKB 4151 Alpha Plus.

LKB 4150 Alpha

This system is a reversed-column chromatograph equipped for pre-column derivatization. The column is made of glass and has solid-state heating. The detection system compromises a dual channel photometer with a high-temperature reaction coil. A single low-volume long-pathlength flow cell is employed. A fluorescence detector is available to provide an approximately 10-fold increase in sensitivity over ninhydrin detection. Refrigerated sample capsule loading is supplied. Powerful programming capability permits the storage of up to 20 methods. Storage facilities for six buffers is supplied.

LKB 4151 Alpha Plus

Alpha Plus is a fully automated and dedicated analyser, this turnkey system has been carefully designed to give a truly robust chromatography. Stepwise elution with up to 5 buffers plus flexible temperature control guarantees optimal separations from even the most complex samples. The versatile programmer monitors and controls all instrument functions and a complete fault-detection system assures absolutely safe operation while preserving the integrity of samples.

The analysis time for protein hydrolysates is 85 minutes using standard columns. For extra high resolution a high-resolution lithium cation exchange column is recommended which achieves baseline separation of virtually all forty amino acids (Figure 6.9).

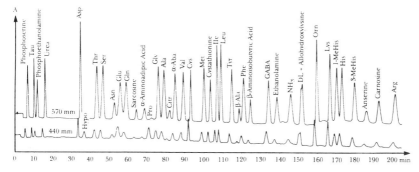

Figure 6.9 *Chromatogram of amino acid mixture: sample physiological fluid amino acid standard, 10 nmol amino acid in 40 µl column – LKB 4418-550 high-resolution column, 270 × 4.6 mm packed with lithium from cation exchange resin; buffers – physiological fluid buffer system (5 lithium buffers); flow rate – 20 ml h^{-1} buffer, 25 ml h^{-1} ninhydrin; detection – vis 570 nm and 440 nm 1.0 AUFS*

6.10 Luminescence and fluorescence spectroscopy

Luminescence is the generic name used to cover all forms of light emission other than that arising from elevated temperature (thermoluminescence). The emission of light through the absorption of UV or visible energy is called photoluminescence, and that caused by chemical reactions is called chemiluminescence. Light emission through the use of enzymes in living systems is called bioluminescence. Photoluminescence may be further subdivided into fluorescence, which is the immediate release (10^{-8} s) of absorbed light energy as opposed to phosphorescence which is the delayed release (10^{-6}–10^2 s) of absorbed light energy.

The excitation spectrum of a molecule is similar to its absorption spectrum while the fluorescence and phosphorescence emission occur at longer wavelengths than the absorbed light. The intensity of the emitted light allows quantitative measurement since, for dilute solutions, the emitted intensity is proportional to concentration. The excitation and emission spectra are characteristic of the molecule and allow qualitative measurements to be made. The inherent advantages of the techniques, particularly fluorescence, are

1 sensitivity, picogram quantities of luminescent materials are frequently studied
2 selectivity, derived from the two characteristic wavelengths and

3 the variety of sampling methods that are available, i.e. dilute and concentrated samples, suspensions, solids, surfaces and combination with chromatographic methods, such as, for example is used in the HPLC separation of *o*-phthalyl dialdehyde derivatized amino acids in natural and sea water samples.

Fluorescence spectroscopy forms the majority of luminescence analyses. However, the recent developments in instrumentation and room-temperature phosphorescence techniques have given rise to practical and fundamental advances which should increase the use of phosphorescence spectroscopy. The sensitivity of phosphorescence is comparable to that of fluorescence and complements the latter by offering a wider range of molecules for study.

The pulsed xenon lamp forms the basis for both fluorescence and phosphorescence measurement. The lamp has a pulse duration at half peak height of 10 µs. Fluorescence is measured at the instant of the flash. Phosphorescence is measured by delaying the time of measurement until the pulse has decayed to zero.

Several methods are employed to allow the observation of phosphorescence. One of the most common techniques is to supercool solutions to a rigid glass state, usually at the temperature of liquid nitrogen (77 K). At these temperatures molecular collisions are greatly reduced and strong phosphorescence signals are observed.

Under certain conditions phosphorescence can be observed at room temperature from organic molecules adsorbed on solid supports such as filter paper, silica and other chromatographic supports.

Phosphorescence can also be detected when the phosphor is incorporated into an ionic micelle. Deoxygenation is still required either by degassing with nitrogen or by the addition of sodium sulfite. Micelle-stabilized room-temperature phosphorescence (MS RTP) promises to be a useful analytical tool for determining a wide variety of compounds such as pesticides and polyaromatic hydrocarbons.

6.10.1 Instrumentation

Luminescence spectrometers

Perkin-Elmer and Hamilton both supply luminescence instruments (see Appendix).

Perkin-Elmer LS-3B and LS-5B luminescence spectrometers
The LS-3B is a fluroescence spectrometer with separate scanning monochromators for excitation and emission, and digital displays of both monochromator wavelengths and signal intensity. The LS-5B is a ratioing

luminescence spectrometer with the capability of measuring fluoroescence, phosphorescence and bio- and chemiluminescence. Delay time (t_d) and gate width (t_g) are variable via the keypad in 10 µs intervals. It corrects excitation and emission spectra.

Both instruments are equipped with a xenon discharge lamp source and have an excitation wavelength range of 230–720 nm and an emission wavelength range of 250–800 nm.

These instruments feature keyboard entry of instrument parameters which combined with digital displays, simplify instrument operation. A high-output pulsed xenon lamp, having low power consumption and minimal ozone production, is incorporated within the optical module.

Through the use of an RS 232C interface, both instruments may be connected to Perkin-Elmer computers for instrument control and external data manipulation.

With the LS-5B instrument, the printing of the sample photomultiplier can be delayed so that it no longer coincides with the flash. When used in this mode, the instrument measures phosphorescence signals. Both the delay of the start of the gate (t_d) and the duration of the gate (t_g) can be selected in multiples of 10 µs from the keyboard. Delay times may be accurately measured, by varying the delay time and noting the intensity at each value.

Specificity in luminescence spectroscopy is achieved because each compound is characterized by an excitation and emission wavelength. The identification of individual compounds is made difficult in complex mixtures because of the lack of structure from conventional excitation or emission spectra. However, by collecting emission on excitation spectra for each increment of the other, a fingerprint of the mixture can be obtained. This is visualized in the form of a time-dimensional contour plot on a three-dimensional isometric plot.

Fluorescence spectrometers are equivalent in their performance to single-beam UV–visible spectrometers in that the spectra they produce are affected by solvent background and the optical characteristics of the instrument. These effects can be overcome by using software built into the Perkin-Elmer LS-5B instrument or by using application software for use with the Perkin-Elmer models 3700 and 7700 computers.

Perkin-Elmer LS-2B microfilter fluorimeter

The model LS-2B is a low-cost, easy-to-operate, filter fluorimeter that scans emission spectra over the wavelength range 390–700 nm (scanning) or 220–650 nm (individual interference filters).

The essentials of a filter fluorimeter are as follows:

- a source of UV/visible energy (pulsed Xenon)
- a method of isolating the excitation wavelength

Table 6.6. *Determination of polyaromatic hydrocarbons in water*

Type of water sample	Compounds named	Comments	Reference
Natural	Benzo(a)pyrene		Shalz and Attman (1968)
Natural	Benzene naphthaline anthracene pyrene fluoranthrene benzo(e)pyrene		Schwarz and Wazik (1976)
Potable	Benzo(a)pyrene		Muel and Lacrox (1960) Jager and Fassovitzova (1968)
Waste	Benzo(a)pyrene	Low temperature	Khesina and Petrova (1973)
Natural	Benzo(a)pyrene	Low temperature	Monarca et al. (1974)

Sewage effluents	Benzo(a)pyrene		Stepanova *et al.* (1972)
Potable	Fluoranthrene benzo(d)fluoranthrene benzo(k)fluoranthrene benzo(a)pyrene benzo(g, h, i)perlyene indeno(1, 2, 3, e, d)pyrene	Low temperature	World Health Organization (1971) Cathrone and Fielding (1978)
Potable	7 H benzo(d, e)anthracene7-one (benzanthrone) fluoranthrene benzo(a)pyrene benzo(g, h, i)pyrene perylene		Ogan *et al.* (1978)
Marine sediments, organisms	Benzo(a)pyrene		Dunn and Stich (1976) Kunte (1967)

- a means of discriminating between fluorescence emission and excitation energy
- a sensitive detector and a display of the fluorescence intensity.

The model LS-2B has all of these features arranged to optimize sensitivity for microsamples. It can also be connected to a highly sensitive 7 µl liquid chromatographic detector for detecting the constituents in the column effluent. It has the capability of measuring fluorescence, time-resolved fluorescence, and bio- and chemiluminescent signals. A 40-portion autosampler is provided. An excitation filter kit containing six filters – 310, 340, 375, 400, 450 and 480 nm – is available to enable the following assays to be performed: fluorescamine, *o*-phthaldialdehyde, 4-methyl-umbelliferone, porphyrins, dansyl derivatives, fluorescein enropium and terbium organo-chelates.

6.10.2 Applications

Potentially, fluorimetry is valuable in every laboratory including water laboratories, performing chemical analyses where the prime requirements are selectivity and sensitivity. While only 5–10% of all molecules possess a native fluorescence, many can be induced to fluoresce by chemical modification or tagged with a fluorescent module.

Hydrocarbons in water

Petroleum products
Petroleum products contain many fluorescing compounds, e.g. aromatic hydrocarbons, polycyclic aromatic hydrocarbons and various heterocylic compounds. Fluorescence techniques have been used to measure these substances in the marine environment (Freegorde 1971; Thruston and Knight 1971; Hargrave and Phillips 1975), spring water (Leoy 1971), natural waters (Lloyd 1971a–c; Parker and Barnes 1960; US Environmental Protection Agency 1979; Danyl and Nietsch 1952; Nietsch 1954, 1956a,b; Leoy 1971), lacustrine sediments (Wakeham 1977; Hargrave and Phillips 1975) and industrial effluents (Dadashev and Agamirova 1957; Rychkova 1969; Aleksova and Gd'dina 1950; Yudilevich 1954, 1960, 1964, 1966; Kurge 1959; Leger, French Patent I, 560, 544; Leonchenkova 1960; Shkylar 1960; Pochkin 1968). Usually the hydrocarbon is extracted from the aqueous sample with a small volume of organic solvent, e.g. benzene, petroleum ether (Dadashev and Agamirova 1957; Yudilevich 1966; Kurge 1959), chloroform, diethyl ether or toluene (Leonchenkova 1960) or non-fluorescent gasoline (Yudilevich 1966).

Polyaromatic hydrocarbons
Some applications of fluorescence spectroscopy to the determination of polyaromatic hydrocarbons in water samples are summarized in Table 6.6. Generally speaking, concentrations down to the picogram ($\mu g\,l^{-1}$), level can be determined by this technique with recovery efficiencies near 100%.

A spectrofluorimeter has been used as a detector in the high-performance liquid chromatographic separation of polyaromatic hydrocarbons in water samples (Vaughan *et al.* 1973; Lewis 1975; Sorrell and Reding 1979; Sorrell *et al.* 1978; Das and Thomas 1978; Schonmann and Kern 1981). A great improvement in sensitivity and specificity can be obtained by the correct choice of wavelengths.

Other applications

Other applications of these techniques in water analysis include the determination of polychlorobiphenyls, chlorophyll measurements, the tracing of fluorescent components in effluent streams, the rate of dilution of pollutant studies and time of travel studies. Fluorescence techniques have also been used for the determination of anions and cations (e.g. cyanide, fluoride, sulphate, phosphates, aluminium, arsenic, beryllium, boron, calcium, magnesium, rare earths, selenium and uranium). For selenium and other inorganics such as boron, aluminium and beryllium, fluorescence offers better sensitivity than atomic absorption.

6.11 NMR spectroscopy

This technique has, to date, found little or no application in the water industry. It is at its most useful in organic structure identification. It also has numerous analytical applications in the fields of polymers and pharmaceuticals. Instrument suppliers are listed in the Appendix.

7 Biological laboratory

7.1 Miscellaneous biological laboratory ware

Naturally in the case of some of the more basic pieces of equipment such as autoclaves, microcentrifuges, refrigerators, freezers, air-flow cabinets, controlled environment rooms, incubators, ovens, microscopes, micropipettes, dispensers, many suppliers exist. Further information on suppliers is available in the Appendix.

7.2 Miscellaneous techniques

7.2.1 Specific spectral techniques

Techniques such as high-performance liquid chromatography and protein and amino acid analysis of interest to the biologist are discussed in earlier sections as is the application of flow-injection analysis for the analysis of carbohydrates and the application of segmented flow analysis to the determination of amino acids.

Fluorescence analysis

Potentially, fluorimetry is valuable in every laboratory performing biochemical analysis and indeed is a routine tool. While fluorescence offers excellent sensitivity (for example the Perkin-Elmer model LS-2B can detect 5×10^{-13} M of fluorescein) the technique is also used for its specific nature, thus simplifying existing methodologies.

Application of the Perkin-Elmer LS-2B memofiller fluorimeter

The determination of DNA is one of the more important assays performed in the molecular biology laboratory. Fluorescence spectroscopy is the key to those assays as it is highly sensitive, easy to use and, with the correct choice of dye, can be made specific for DNA even in the presence of RNA and proteins. The fluorimetric assays generally use dyes which produce a large increase in the fluorescence signal on binding to DNA. The most common of these dyes are ethidium bromide, Hoechst 33258 and 4,6-diamido-phenyl indole (DAPI).

The Hoechst 33258 DNA assay method is approximately 500–10 000-fold more sensitive than classical UV spectroscopic and diphenylamine colorimetric methods respectively. The dye may also be used in conjunction with ethidium bromide to determine the (G + C) content of DNA of unknown base composition.

In the DAPI method DNA is assayed quantitatively with excitation at 240 nm and emission at 466 nm.

Below are discussed some of the applications of the Perkin-Elmer LS-2B micrifiller fluorimeter in biochemical assay work.

In time-drive mode, the model LS-2B can be used for monitoring rate-determining reactions. For example, glucose can be measured in solution using an assay procedure based upon the following reaction:

$$\text{D-glucose} + O_2 + H_2O \xrightarrow{\text{GOD}} \text{D-gluconic acid} + H_2O_2$$

$$H_2O_2 + \text{homovanillic acid} \xrightarrow{\text{POD}} \text{fluorescent product} + 2H_2O$$

At low substrate levels, the rate of the reaction rises linearly with concentration. Thus by monitoring the fluorescence emission of the end-product with respect to time at varying glucose concentrations, a calibration curve can be constructed of concentration versus rate. The fluorescent product is excited at 310 nm and the emission measured at 440 nm.

Application of the Hamilton chemi-Lumicon bioluminscence luminometer

An example of the application of this system is the measurement of adenine triphosphate (ATP), using the firefly luciferin–luciferase luminescence system. The Lumicon can be used in two modes: the peak mode for fast decaying pulses of light (<10 s) and the repeat mode for slow kinetics (up to 100 minutes) such as the luciferin–luciferase + ADP system. The repeat mode presents observation windows (or counting intervals) between 1 and 99 seconds and repetitive measurements to be performed up to 99 minutes. Three solution injectors can be activated and various other parameters such as delay time after injection, real time, blank subtraction, background subtraction and factorization of the counts are set by the user. Injector parameters such as dispense speed, a different volume for each position if required and the time of injection during the experiment are user-definable.

Eppendorf Analysor POS 5060 sective spectrophotometric analysis system

This is a visual spectrophotometric system designed specifically for use in the medical laboratory but may have applications in the water biology laboratory.

The system analyser 5060-data terminal 6410 is capable of carrying out 300 analyses per hour. 30 parameters are operator selectable. They include kinetic enzyme activity measurements, aldolase, AP, amylase, CHE, CK, CK-MB, GLDH, GOT (ASAT), GPT (ALAT), GGT, HBDH, LAP, LDH, lipase, acid phosphatase (kinetic substrate concentration determinations). Glucose (GOD-PAP, Glucoquant, Gluc DH), urea, uric acid, creatinine, triglycerides (enzyme-immunoassays), carbamazepine, ethosuximide, gentamicin, phenobarbital, phenytoin, primidone, theophylline, tobramycin, thyroxine, valproic acid (turbidimetric measurements), antithrombin, heparin, IGA, IGG, IGH, transferrin, albumin, C-3 complement, C-4 complement, caeruloplasmin, haptoglobuline and a 2-macroglobuline and (end-point determinations) alcohol, inorganic phosphorus, bilirubin, calcium, chloride, cholesterol, iron, total protein, uric acid, lactate and triglycerides.

Carbon dioxide analysers

This instrument is for the measurement of carbon dioxide in incubator gases or headspace gases in reaction vessels. The Ciba Geigy 965 D carbon dioxide analyser supplied by Denley in the UK is based on the Van Slyke method and covers the concentration range $0-12\,g\,l^{-1}$ carbon dioxide.

7.2.2 Biosensors

Biosensors employ biological species (such as enzymes, antibodies, etc.) to give reactions measurable by an appropriate transducer (i.e. a device to interpret a chemical reaction as an electrical signal, such as an electrode or a photodetector). The biological species is closely associated with the transducer surface.

An array of non-electrochemical phenomena have been used in the fabrication of biosensor devices, particularly those based on optics, such as surface plasma resonance, evanescent wave techniques (total internal reflection spectroscopy) and ellipsometry. These optical approaches have in common the analysis of surface changes following the binding of biological species and as such are particularly suitable for immunoassay (discussed later). Other approaches have been the use of thermometry, conductimetry and piezoelectric crystals.

Enzyme/electrode combinations have been studied more than any other type of biosensor, with a considerable proportion of the literature being directed to glucose and urea. In fact, the enzyme electrode is the only one that has yet reached the market place. Cambridge Life Sciences plc is one supplier of biosensors.

A wide range of other substances have been looked at including amino acids, antibiotics, phenols, alcohols, carbohydrates, triglycerides, urea and uric acid.

7.2.3 Virus and bacteria removal

The removal of viruses or bacteria from water samples or their recovery from a water sample on a filter for further examination are an important feature of work carried out in a biology laboratory. The Anopore range of filters is capable of removing bacteria and viruses from water – see also Chapter 3.

7.3 Bioassays/immunoassays

Biology and biochemistry laboratories perform three general types of assays:

1 Binding assays including the following: immunoassays such as radio immunoassay (RIA), fluorescence immunoassay (FIA), enzyme immunoassay (EIO), enzyme-linked immunoassay (ELISA and EMIT)
2 Enzyme assays – both kinetic and end-point radiocoordination of proteins, lipid assays, receptor binding assays and tissue-culture techniques
3 Chemical assays such as total protein assays and analytical chemistry including spectroscopy and chromatography.

Assays 1 and 2 are described below. Assay type 3 is discussed earlier under specific chemical techniques.

Immunoassays

Immunoassays (type 2) are based on the following reaction:

$$\underset{\text{antigen}}{[Ag]} + \underset{\text{antibody}}{[Ab]} \rightleftharpoons \underset{\text{complex}}{[Ab - Ag]}$$

Each of the types of immunoassay listed above (RIA, FIA, EIA, ELISA and EMIT) has its own advantages. In general immunoassays involve large numbers of samples and are a source of routine, repetitive work. Whatever the type, immunoassays require the following equipment:

- liquid handling – pipetting, dispensing, etc.
- sample conditioning – mixing, incubating, etc.
- separation – centrifugation, filtration, etc.
- measurement – spectroscopy, gamma counter, etc.

Immunoassays are nowadays performed by one of two approaches either partially automated or fully automated (robotics).

7.3.1 Partially automated immunoassay systems

The separate items of equipment necessary for the preliminary (i.e. sample preparation) stages of partially automated (i.e. prior to the final measurement instrument) immunoassay available from Denby Instruments Ltd are listed in the Appendix.

Having completed the sample preparation stages the final measurement can be made by a variety of techniques.

Luminescence and fluroescence measurements in luminescent immunoassays

Perkin-Elmer LS-2B microfilter fluorimeter
Fluorescence is used in immunochemistry. Essentially the radioactive tag on the antigen is replaced by a fluorophore. The most commonly used tags are fluorescein and umbelliferone.

Organic chelates of certain lanthanides such as Tb^{3+} and Eu^{3+} are used as a means of removing unwanted background fluorescence in fluoroimmunoassays. The lanthanide chelates exhibit long-lived emission 50 µs to 3 ms. Using a gated detection system such as that of the model LS-2B the non-specific background fluorescence with a decay time of 100 ns is discriminated against.

Hamilton Umicon Lumicon chemi- and biolumium assay luminometer
This equipment is used in test-tube scale luminescent immunoassays. With its sample compartment (thermostatted by means of Pettier elements, which allow the temperature to be set from 15°C to 40°C with a precision of 0.1 °K) this instrument is suitable for the measurement of temperature-sensitive bioluminscence resulting from enzymic reactions and also in phagocyte-mediated luminescence measurements.

This instrument can be used in two modes: the peak mode for fast-decaying pulses of light (<10 s) and the repeat mode for slow kinetics (up to 100 min).

Spectrophotometric plate readers
Perkin-Elmer's lambda reader, an automated microprocessor-controlled, microplate reader, offers the flexibility of configuring a reliable, user-friendly, versatile system, capable of accommodating a wide variety of assays requiring colorimetric measurement on microscale (<300 µl) samples. These assays include ELISA, protein determination, cytotoxicity, cytoproliferation and antibody sensitivity testing.

Radio-immunoassay (RIA) analysers
Kontron instruments supply the MDA 312 multi-detector RIA analyser. This is the first multi-detector gamma counter incorporating a multi-channel

instead of a conventional pulse-height analyser. During the counting period the pulse-height spectra of each individual sample (detector) is recorded and stored. Starting from the photopeak the MDA 312 now sets the energy-window levels on both sides of the peak. This way the counting window is automatically optimized for each sample.

The direct benefits of this innovative technique are twofold: the counting efficiency of an individual detector is optimal and constant.

The possibility of centring the photopeak into the counting window for each individual sample allows the selection of a relatively narrow window, whereas in the conventional technique the windows have to be set much broader in order to compensate for drifts. The result is a substantial reduction in background count rates.

Packard supply the Cobra – one Auto-gamma 5012 and 5013 instruments. This combines RiaSmart data reduction, expert quality control management and high-energy counting capabilities. It has been used in ^{125}I radioimmunoassay measurement and DNA probe analysis using ^{32}P. It features three simultaneous counting regions, half-life correction, spill-over compensation, spectral display and plotting and an on-board computer. This company also supplies the Crystal + plus benchtop manual RIA system. This is available in 24-, 12- or economical 6-detector configurations: the 24-detector model can count over 1400 samples per hour. A built-in multi-tasking microcomputer saves time by simultaneously counting samples and reporting results and the Crystal + plus can connect to a labmicro or mainframe for more extensive data reduction.

Up to 50 stored assay protocols include routines for RIA/IRMA, dual label assays, T3-uptake and FTI calculations and hepatitis, RAST and hCG screening. Quality-control charts can be printed automatically.

Biosensors
Biosensors are used in the final measurement stage of immunoassays such as an electrochemistry-based enzyme-linked immunoassay (ELISA) or the measurement of catalyse-labelled antigens at an antibody-coated oxygen electrode.

7.3.2 Fully automated immunoassay systems

The following benefits accrue from full automation:

1 improved precision by reducing human errors
2 freeing sensor personnel from repetitive tasks
3 isolating personnel from hazardous environments and protecting experiments from human contamination
4 faster sample turn-around

The Zymark robotic laboratory automation system

Although detail procedures differ in each laboratory, the basic elements of binding and enzyme assays are similar. The generalized procedure shown in Table 7.1 highlights the common steps and indicates which Zymate laboratory systems are required. These procedures are performed using common laboratory glassware such as test tubes or in multiple tube devices such as microtitre plates.

Table 7.1. *Typical immunoassay procedure using Zymate robotic laboratory automation system*

Step	Procedure and comment	Zymate laboratory stations
1	*Dispense samples, standards and controls* Precision dispensing or pipetting of samples, standards and controls.	Z510 Master Laboratory Station, or Z910 Precision Microlitre syringe head
2	*Serial dilute – as required* Precision dispensing of reagents or buffers then transferring an aliquot from the previous dilution.	Z510 Master Laboratory Station, or Z910 Precision Microlitre syringe head
3	*Mix* Votex if possible. Vortexing is gentle, simple to perform and eliminates cross contamination.	Z620 Vortex Station
4	*Transfer samples if required* Transfer sample aliquots to an appropriate tube or container.	Z910 Precision Microlitre syringe head with pipette tips
5	*Incubate* Hold samples at a programmed temperature and time.	Zymate controller and Z830 Power and event controller
6	*Quench or terminate reaction* Dispense reagent or buffer to terminate reaction.	Z510 Master laboratory station
7	*Separate bound and free reactants* The following alternatives are available: A Filtration – retaining either the filtrate or retentate	Several options available

Table 7.1. (*continued*)

Step	Procedure and comment	Zymate laboratory stations
	B Centrifugation	Centrifuge station 1Q84
	C Adsorption using solid phase extraction minicolumns	Z510 Master Laboratory Station and Z911 Liquid distribution hand
	D Immobilized substrate where material to be removed is bound to a large substrate permitting simple mechanical separation	Robotic manipulation
8	*Wash and Dry* Dispense and remove wash reagents and, if needed, control gas purge.	Z510 Master Laboratory Station
9	Add scintillation cocktail or other detection amplifying reagent.	Z510 Master Laboratory Station
10	Introduce sample into measurement instrument and monitor results.	Z830 Power and event control station. 7820 Printer

Beckman Robotic Biomek 1000 automated laboratory workstation

The Biomek 1000 integrates the work formerly done by four instruments: sample preparation system, diluter/dispenser, plate washer and a spectrometer finish. In can handle assays such as radio-immunoassays (RIA), fluorescence immunoassays (FIA) enzyme immunoassays EIA and enzyme-linked immunoassays (ELISA).

7.4 Analysis of solid materials of biological interest

The water laboratory is frequently requested to determine trace elements in biological materials such as fish, crustacea and plant material. Such measurements are used as markers for the amount of pick-up of these elements from the aqueous environment over a period of time, which is useful alternative information to that provided by spot measurements on water samples.

The sample when it arrives in the laboratory is usually in a form unsuitable for analysis such as deep frozen wet fish, plant material, a river sediment or

Table 7.2.

Fish, crustacea	Sediments, sludges	Plant material
Homogenize wet sample	Comminute sieve	Comminute digest sample with acid
Determine moisture content	Particle-size measurement possibly	
Digest sample with acid	Digest sample with acid	

sewage sludge suspended in water. Table 7.2 shows steps that may then be required to convert the sample into a form suitable for analysis.

Note If determinations of certain volatile elements such as mercury or selenium are required it is necessary to carry out these analyses on the wet sample as received (to avoid loss of element by drying at 105°C). The dry weight of material in the sample is obtained by determining moisture in a separate position of the sample and applying a correction to the sample weight used in metals determination.

As an alternative to drying at 105°C microwave drying has been used (Kuchn *et al.* 1986) to remove moisture from aqueous slurries.

7.4.1 Homogenization of fish samples

Portions of the deep frozen fish sample are sampled statistically, combined and allowed to thaw. It is now necessary to homogenize this material to obtain a homogeneous sample by chopping with a stainless steel knife or alternatively with a vibration micropulverizer (Table 7.3B) or a rotor speed mill (Table 7.3C). Alternatively, the deep frozen sample can be mixed with cardice and ground in a planetary micromill (Table 7.3A). Using either method is is necessary to check that no trace element contamination occurs from the homogenizing equipment used. The wet slurry thus obtained is either analysed direct in the wet condition or, if volatile metals are not being determined, analysed after drying to constant weight at 105°C.

7.4.2 Comminution of sediments and plant material

Various comminution devices (Table 7.3D–H) are available for handling these types of samples.

Grinding elements are offered in various non-contaminating materials such as corundum (Al_2O_3), agate (SiO_2), or zirconium oxide (ZrO_2).

Table 7.3. *Laboratory homogenizers and comminution equipment supplied by Fritsch*

	Homogenizers	Description	Sample type
A	Planetary micromill	Pulverisette 7	Homogenizing wet fish or crustacea pastes or grinding to 0.1 μm deep-frozen fish or crustacea
B	Vibration micropulverizer	Pulverisette 10	Homogenization of dry or wet fish or crustacea samples, sample preparation in microbiology
C	Rotar speed mill	Pulverisette 14	Grinding to 50 μm of organic materials, e.g. plant materials
Comminution			
D	Vibrating cup mill	Pulverisette 9	A gate grinding to 20 μm of dry or wet sediments, fish, crustacea or plant material
E	Laboratory desk mill	Pulverisette 13	Grinding 0.1 mm of dried sediments, soils, sewage sludge, hydrological sediments and drilling cores
F	Mortar grinder	Pulverisette 2	Grinding to 10 μm of sediments and soils
G	Centrifugal mill	Pulverisette 6	Grinding to 1 μm of sediments and soils
H	Planetary mill	Pulverisette 5	Grinding to 0.1 μm of sediments and soils
I	Sieving devices Vibratory sieve Shaker for micro-precision sieving	Analysette 3	Dry and wet sieving (25 mm to 20 μm) or micro-precision sieving (5–100 μm)
J	Rotary sieve shaker	Analysette 18	Satisfies ASTM E-11-190 BS 410 1969; AFNOR NFX 11-501 and DIN 53477/1 separation of coarse grain material

7.4.3 Sieving analysis of dry sediments

Having comminuted the sample it may now be required to carry out a sieving analysis in order to obtain different size fractions for chemical analysis. Fritsche supply a range of devices for sieving analysers (Table 7.3I,J).

7.4.4 Particle-size distribution measurement

A complete particle-size analysis can require the use of various analysis technologies. A microscopic examination may be performed before the sieve analysis, which in turn can be followed by a sedimentation analysis or the recording and the evaluation of a diffraction pattern.

The working ranges of the analysis methods overlap and can be subdivided as shown in Table 7.4, which also details equipment suppliers (see also the Appendix).

Sieving methods (5 μm to 63 mm)

Sieving methods have been discussed in the previous section.

Gravitational sedimentation, 0.5–500 μm

An optical measuring system is used in sedimentation analysis, whereby a concentrated beam of light is deflected horizontally through the lower section of a measuring vessel onto a photoelectric cell. The amount of light absorbed by the sedimenting particles decreases with time as the number of particles passing the measuring beam increases. The increase in the photoelectric current as a function of time is then a measure of the particle size.

A major step on the road to reducing the measuring time is provided by the 'Analysette 20' scanning photo sedimentograph (Table 7.3). Using this device, the measuring time is considerably reduced by a continuous movement of the light beam towards the direction of fall of the particle.

Centrifugal sedimentation (0.05–10 μm)

The Andereasen pipette (Fritsch Analysette 21) is extremely well suited to this type of analysis. The measuring radius is determined by six rotating capillaries of equal length in a centrifuge drum. At certain predetermined times, samples are drawn from this radius using a pipette and the solid content of these samples mathematically evaluated to determine the particle-size distribution for the whole sample. The volume of material remaining in the centrifuge is reduced with each sampling and the distance

Table 7.4. *Suppliers and working ranges of particle-size distribution methods*

Method	Particle-size range	Equipment supplier	Model
Dry sieving	63um–63 mm	Fritsch	Analysette 3, (20 µm–25 mm)
Wet sieving	20 µm–200 µm	Fritsch	Analysette 18, see Table 7.3 A–C
Microsieving	5 µm–100 µm	Fritsch	
Sedimentation in gravitational field	0.5 µm–500 µm	Fritsch	Analysette 20
Laser diffraction	0.1 µm–1100 µm	Fritsch	Analysette 22
Electrical zone sensing	0.4 µm–1200 µm	Coulter	Model ZM, Coulter multisizer
Electron microscopy	0.5 µm–100 µm	–	–
Photocorrelation spectroscopy	0.5 µm–5 µm	–	–
Sedimentation in centrifugal field	0.5 µm–10 µm	Fritsch	Analysette 21 (Anderson Pipette centrifuge)
Diffraction spectroscopy	1 µm–1 mm		
Optical microscale	0.5 µm–1 mm		
Projection microscopy	0.05 µm–1 mm		
Image analysis systems	0.8 µm–150 µm down to 0.5 µm	Joyce–Leebl Leitz, Karl Zeiss, Cambridge Instruments	Magiscan and Magiscan P Autoscope P Videoplan II Quantimet 520

between the surface of the sample liquid and the measuring plane is also reduced, thus reducing the sedimentation time for the smallest particles without the accuracy of the measurement being affected.

Laser diffraction (0.1–1100 µm)

This is a universally applicable instrument for determining particle-size distributions of all kinds of solids which can be analysed either in suspension in a measuring cell or dry by feeding through a solid particle feeder. In the Fritsch Analysette 22 laser diffraction apparatus the measured particle-size distribution is displayed on the monitor in various forms, either as a frequency distribution, as a summary curve or in tabular form and can be subsequently recorded on a plotter, stored on hard disk or transferred to a central computer via an interface. The time required for one measurement is approximately 2 minutes.

Electrical zone sensing (0.4–1200 µm)

This is the classical method of carrying out particle-size analysis. Coulter supply two instruments – the Model ZM (video display optical) and the top-of-the-range multisizer – the latter having built-in video display of results.

The Coulter method of sizing and counting is based on measurable changes in electrical resistance produced by non-conductive particles suspended in an electrolyte.

By means of the Coulter channelizer 256 module an optional extra on the model ZM but built-in on the multisizer, enables biological cell-size distributions to be measured. This provides an ability to measure suspension concentration and distributions of populations against size with a choice of 64-, 128- or 256-channel resolution over a range approximately 3:1 diameter. Size differences as small as $0.05\,\mu m^3$ (fL) are detected.

A data management system is also available for the model ZM.

7.4.5 Digestion of solid samples preparatory to chemical analysis

Having as necessary, dried, homogenized or comminuted the samples, they must now be digested in a suitable reagent to extract elements in a suitable form for chemical analysis. In many organizations we have reached the point where the analyses pass from the hands of the biologist to those of the analytical chemist. In the author's experience, however, it must be emphasized that to ensure best-quality results the whole procedure from, for example, statistically sampling a fish to the final chemical analysis, should be handled by the same person.

Wet ashing

Digestion of the sample with hydrochloric acid, hydrofluoric and (if silicaceous material present) nitric acid and aqua regia have all been used. Aqua regia will dissolve most metals. Nitric acid provides an oxidizing attack for organic materials which are usually present at very high concentrations in biological specimens. Perchloric acid is a very strong oxidizing agent, especially when used in conjunction with nitric acid, but its use is not favoured by all chemists and certainly it must not be used in the pressure dissolution technique discussed below.

Fusion

Fusion with a flux such as sodium hydroxide, potassium bifluoride and potassium pyrosulphate has been used extensively in the water industry.

Dry ashing

This is often used to remove organic material from biological material. The sample is weighed into a suitable container such as a ceramic or metal crucible, heated in a muffle furnace and the residue dissolved in an appropriate acid. It is not suitable for the analysis of volatile elements such as mercury and arsenic, since they may volatize during the ashing process. Magnesium nitrate has been used as an ashing agent to prevent volatilization or arsenic during dry ashing. Dry ashing has been used in the analysis of plant material and municipal waste (Haynes 1978; Dalton and Melanoski 1969).

Pressure dissolution

Pressure dissolution and digestion bombs have been used to dissolve samples for which wet digestion is unsuitable. In this technique the sample is placed in a pressure dissolution vessel with a suitable mixture of acids and the combination of temperature and pressure effects dissolution of the sample. This technique is particularly useful for the analysis of volatile elements which may be lost in an open digestion. Pressure dissolution has been used for the determination of volatiles in fish, bird and plant tissue (Adrian 1971).

Microwave dissolution

More recently, microwave ovens have been used for sample dissolution. The sample is sealed in a teflon bottle or a specially designed microwave digestion vessel with a mixture of suitable acids. The high-frequency microwave, temperature (ca. 100–250°C) and increased pressure have a role to play in the success of this technique. An added advantage is the significant reduction in sample dissolution time (Raverz and Hasty 1987). Nadkarni (1984) used microwave dissolution in the analysis of biological samples and sediments.

Equipment for sample digestions

Pressure dissolution acid digestion bombs

Inorganic and organic materials can be dissolved rapidly in Parr acid digestion bombs with teflon liners and using strong mineral acids, usually nitric and/or aqua regia and, occasionally, hydrofluoric acid. Perchloric acid must not be used in these bombs due to the high risk of explosion.

Table 7.5 contains temperature and pressure data obtained while using microwave heating with a single closed vessel for two different acids. For nitric acid, 200°C (80°C over the atmospheric boiling point) and $100\,lb\,in^{-2}$ was achieved in 12 minutes and for hydrochloric acid 153°C (43°C over the atmospheric boiling point) and $100\,lb\,in^{-2}$ was obtained in 5 minutes.

At such elevated temperatures these and other acids become more corrosive. Materials that digest slowly or will not digest at the atmospheric boiling points of the acids become more soluble so dissolution times are greatly reduced. The aggressive digestion action produced at the higher

Table 7.5. *Temperature pressure data for acids heated in a 120 ml closed vessel*

Acid (wt%)		Temperature (°C)	Pressure (lb in^{-2})
HNO$_3$	70	200	120
HCl	37	153	120
HNO$_3$	70	193	100
HCl	37	130	100

Table 7.6. *Single-vessel dissolution of inorganic sample using HF:HNO$_3$:H$_2$O*

Sample size	Acid volume	Microwave digestion time	Hot-plate digestion time	Time saved
1 g	36 ml[1]	1 hour	5 hours	4 hours

[1] HF:HNO$_3$:H$_2$O, 1:1:1
HF – 48 wt. %
HNO$_3$ – 70 wt. %

temperatures and pressures generated in these bombs result in remarkably short digestion times, with many materials requiring less than one minute to obtain a complete dissolution, i.e. considerably quicker than open-tube wet-ashing or acid-digestion procedures (Table 7.6).

Kingston and Jassie (1986) have studied the use of microwave energy for the acid decomposition at elevated temperatures and pressures of oyster tissue and organs.

Several manufacturers supply microwave ovens and digestion bombs (Tables 7.7 and 7.8). CFM Corporation state that their solid PTFE bombs are suitable for the digestion of fish and plant tissues.

Oxygen combustion bombs (Table 7.7)

Combustion with oxygen in a sealed Parr bomb has been accepted for many years as a standard method for converting solid and liquid combustible samples into soluble forms for chemical analysis. It is a reliable method whose effectiveness stems from its ability to treat samples quickly and conveniently within a closed system without losing any of the sample or its

Table 7.7. *Microwave oven digestion bombs*

Supplier	*Oven part no.*	*Bomb part no.*	*Comments*
Acid digestor types			
Parr Instruments	Not supplied	4781 4782	See Table 7.8 for metals determination in organic material
CEM Corporation	MD581D	Solid PTFE	See Table 7.8 for metals determination in organic material
Prolabo	Microdigest 300 Microdigest A300	Solid PTFE	See Table 7.8 for metals determination in organic material
Oxygen combustion types			
Parr instruments	Not supplied	1108	For sulphur, chlorine, etc. determination in biological material

Table 7.8. *Microwave digestion bombs supplied by Barr Instruments*

Catalogue No.	4781	4782
Maximum charge of:		
inorganic sample	1.0	1.0
organic sample	0.1	0.2
Maximum temperature (°C)	250	250
Cup seal	Teflon-o-ring	Teflon-o-ring
Overpressure protection	Compressible relief disc	Compressible relief disc
Closure style	Band tighten	Band tighten
Bomb dimensions, cm:		
height overall	112	14.3
max. o.d.	7.8	7.8
Cup dimensions, cm:		
Inner diameter	3.1	3.1
Inner depth	3.0	6.1
weight, g	515	625

combustion products. Sulphur compounds are converted to soluble forms and absorbed in a small amount of water placed in the bomb. Organic chlorine compounds are converted to hydrochloric acid or chlorides. Any mineral constituents remain as ash but other inorganic elements such as arsenic, boron, mercury, phosphorus and nitrogen and all of the halogens are recovered with the bomb washings. In recent years the list of applications has been expanded to include metals such as chromium, iron, nickel, manganese, beryllium, cadmium, copper, lead, vanadium and zinc by using a quartz liner to eliminate interference from trace amounts of heavy metals leached from the bomb walls and electrodes (*Parr Manual* (1974); Nadkarni 1981).

7.4.6 Elemental analysis of sample digests

Once the sample is in solution in the acid and the digest made up to a standard volume the determination of metals is completed by standard procedures such as atomic absorption spectrometry or inductively coupled plasma optical emission spectrometry.

If the sample matrix is complex, it may be necessary to determine if there are any interference effects from the matrix, on the analyte response. This is usually done by spiking the sample with a known amount of analyte. Two equal portions of sample are taken and an appropriate quantity of analyte is added to one to effectively double the absorbance. A similar quantity of analyte is added to water to make a 'spike-alone' solution. Readings are taken for sample, sample-plus-spike and spike-alone solutions and the amount of interference calculated as a percentage enhancement or suppression of the response. The interference can then be corrected or preferably removed by use of a separation technique.

It is advisable to include in the sample run standard materials of a type similar to the samples being examined. Standard biological materials and river sediments are available from the National Bureau of Standards USA.

7.5 Non-destructive analysis of biological material

An area of increasing interest is the elemental analysis of biological materials by non-destructive methods such as X-ray fluorescence spectrometry (Section 2.10) and neutron activation analysis (Section 2.11). Although most water laboratories do not have facilities for carrying out these techniques on their own premises, facilities are available on a consultancy basis. Thus, the Atomic Research Establishment at Wantage will carry out neutron activation analysis under contract at a reasonable cost. The Water Research Association, UK, has been investigating the application of X-ray fluorescence spectroscopy to solid samples. Some advantages of non-destructive methods are no risk of loss of elements during sample handling operations, the absence of contamination from reagents, etc. and the avoidance of capital outlay on expensive instruments and highly trained staff.

7.5.1 Applications

In Table 7.9 are shown results obtained in the digestion in closed vessels of 1 g samples of MBS SRM 1645 river sediment samples, digested (a) in 20 ml of 1:1 nitric acid water and (b) in 5 ml concentrated nitric acid and 3 ml 30% hydrogen peroxide. In the former, at a power input of 450 W, the temperature and pressure rose to 180°C and 100 lb in^{-2}. At that point, microwave power was reduced to maintain the temperature and pressure at those values for an additional 50 minutes. In the latter case, one gram samples were open-vessel digested in 1:1 nitric acid: water for 10 minutes at 180 W. After cooling to room temperature, 5 ml of concentrated nitric acid and 3 ml of 30% hydrogen peroxide were added to each. The vessels were then sealed and power was applied for 15 minutes at 180 W followed by 15

Table 7.9. *SRM 1645 river sediment microwave digested in 1:1 HNO_3:H_2O*

	(a) in 1:1 HNO_3:H_2O	(b) in 5:3 HNO_3: H_2O_2	
Element	Amount recovered (%)	Amount recovered (%)	Certified value (%)
			0.0066
As	0.0060, 0.0060	0.0075, 0.0070	0.0012 ± 0.00015
CD	0.0012, 0.0012	0.0011, 0.0012	2.96 ± 0.28
Cr	3.00, 2.98	3.04, 2.96	0.0109 ± 0.0019
Cu	0.0122, 0.0113	0.0118, 0.0119	0.74 ± 0.02
Mg	0.72, 0.72	0.70, 0.70	0.0785 ± 0.0097
Mn	0.0790, 0.0780	0.0720, 0.0725	0.00458 ± 0.00029
Ni	0.0050, 0.0050	0.0044, 0.0044	0.0714 ± 0.0028
Pb	0.0736, 0.0737	0.0736, 0.0733	(0.00015)
Se	0.0001, 0.0001	0.0001, 0.0001	0.172 ± 0.017
Zn	0.170, 0.168	0.160, 0.160	

minutes at 300 W power. As can be seen, the temperature rose to 115°C after the first 15 minutes and to 152°C at 38 lb in^{-2} after the final 15 minutes of heating. With both reagent systems element recoveries are in good agreement with the certified values obtained using a hot plate total sample digestion technique which typically requires 4–6 hours.

In Table 7.10 is demonstrated the fact that in the case of sewage sludge the use of closed vessels in combination with microwave heating can speed up sludge digestion significantly. To demonstrate this, three sets of duplicate samples of the same standard EPA sludge sample were digested using different methods. The first set was microwave digested in closed vessels using 70% nitric acid and 30% hydrogen peroxide. The second set was also microwave digested in closed vessels but 1:1 nitric acid: water was used. The third set of samples was digested in glass beakers on a hot plate following EPA SW-846 procedures. The microwave digestions required 40 minutes for the nitric acid: hydrogen peroxide dissolution and 60 minutes for the 1:1 nitric acid: water dissolution. The hot-plate dissolution required 10 h. Agreement on element recoveries among the three digestion procedures is very good and, except for arsenic, they agree well with the average sample reference values.

Topping (1987) has carried out a detailed study of the determinations of trace metals in biological reference materials, namely fish, crab and lobster. Samples were obtained from the Atlantic, North Sea, Barents Sea and Baltic Sea.

Table 7.10. *Element recovery for closed vessel microwave digestion versus open vessel hot plate digestion of EPA sludge sample (all values are µg g⁻¹)*

Digestion	Ag	As	Cd	Cr	Cu	Fe	Pb	Ni	Se	Zn
Microwave closed vessel										
10 ml HNO_3	81	4	20	184	1139	17,500	575	185	4	1308
3 ml H_2O_2	81	3	20	184	1132	17,450	576	187	4	1308
40 min										
Microwave closed vessel										
20 ml 1:1 HNO_3:H_2O	80	4	21	188	1136	18,350	593	197	4	1319
60 min	78	4	21	187	1125	18,550	598	193	4	1325
Open vessel hot plate										
10 ml HNO_3	84	3	21	183	1117	17,700	601	191	4	1283
4 ml H_2O_2	88	3	21	183	1128	17,650	591	192	4	1287
10 h										
EPA $\bar{X}$ value	80.6	17.0[1]	19.1	193	1080	16,500	526	194	not reported	1320

[1] Range of reported values was 0–89 µg g⁻¹.

Most participants used atomic absorption spectrometry procedures for the measurement of metals in marine tissue. In the first two exercises of the intercomparison programme analysts employed flame techniques for the analysis of copper, zinc, cadmium and lead and cold vapour technique for the analysis of mercury. The introduction of heated graphite furnace attachments in 1974 resulted in an increasing number of analysts employing a flameless technique for the analysis of cadmium and lead and a few used a chelation/extraction procedure to isolate these elements from the matrix. In the fourth exercise analysts were asked to include arsenic in their suite of elements; most analysts employed a hydride-generation technique coupled with either a flame procedure (H_2/N_2 or H_2/Ar) or a flameless procedure utilizing a heated quartz tube. The results of these exercises showed that in the more recent exercises there was a progressive improvement in the comparability of analytical results of copper, zinc and mercury at concentrations normally found in fish and shellfish tissue. However, the results of the cadmium and lead analyses were in general not as encouraging.

7.6 Automated biochemical oxygen demand (BOD)

Manual determinations of BOD is very labour intensive and forms a high proportion of the workload of routine water-testing laboratories. Various instruments have been developed for reducing this workload, ranging from simple labour saving devices to the application of robotics.

Manometer determination of BOD

pH OX Systems supply apparatus for this measurement (models 212 and 214 BOD manometers). The manometric method is much less time consuming than the standard dilution method. The manometric principle of measurement is not affected by the amount of dilution and has a wide measuring range.

In the course of a determination, the bacteria in the sample consume oxygen in the oxidation of organic substances. The oxygen consumed in this way is replaced from the air above the sample; at the same time carbon dioxide is released. This is chemically bonded by a 45% potash lye and is eliminated from the system in this way. This results in a negative pressure, which is read off the scale as mg/l BOD.

When determining BOD manometrically a physical change is measured, but the sample itself is not destroyed. A continuous curve of BOD+ values can be drawn in this way, which provides a great deal of information on the nature of the sample.

Tech-line supply a similar instrument, the model 025-601 waste-water respirometer, and claim that results of short-term (12–24 hours) BOD measurements can be correlated with 5-day BODs and that precision is ±5%.

Skalar automated BOD

The Skalar computer's automatic sampling handling and processing system (Sample Processor 100) can be used to automate the determination of dissolved oxygen and consequently of BOD. This equipment applies robotics to the determination of dissolved oxygen but, unlike the Zymark system, discussed below, does not automate the sampling and incubation stages of BOD measurements.

Zymark Zymate II system

This robotic system is capable of carrying out completely automated BOD measurements. The automated method is based on the standard dissolved oxygen electrode measurement five-day dilution method used by the American Public Health Association and approved by the Environmental Protection Agency.

The system carries out the following operations:

1 *System set up*   Manually transfer incoming samples into standardized containers.
2 *Measure and adjust pH and chloride concentration if required*   Measure the pH using the pH workstation module. Adjustments to the pH can be made using the master laboratory station. Monitor and adjust chlorine concentration using the same modules. Clean the pH and chlorine ion probes using the wash station.
3 *Prepare samples*   Measure and transfer aliquots of samples into standard BOD bottles using the syringe hand. Add seed solution if required. Dispense diluent water to volume using the dissolved oxygen diluent water station.
4 *Measure initial dissolved oxygen level*   Determine the initial dissolved oxygen (DO) using the same station and store the value in the Easylab program. Cap the BOD bottles using the BOD capping station.
5 *Incubate samples*   Manually remove samples from the racks and place in a dark incubator for five days. Return samples manually to the Zymate system work area after incubation.
6 *Measure final (DO) level*   Remove cap from the sample using the BOD capping station and place the bottle in the DO/diluent water work station. Measure final DO. Calculate the BOD using the initial dissolved oxygen reading, the final reading, the dilution used and the correction for the seed stored in the Easylab program. Print results using the printer and transmit to a laboratory computer if required.

Malthus system

This system, devised by the Wessex Water Authority, is based on continuous electrical conductivity measurements. Several hundred samples can be measured simultaneously. As well as monitoring oxygen uptake, this system from the shape of the curve obtained can, in some cases, provide information on the type of bacteria present in the sample.

8 Radioactivity measurements

The production of electricity in many parts of the world is based to an ever-increasing extent upon the growth of the nuclear power industry. This has been accompanied by increasing public concern, heightened by events such as Three Mile Island and Chernobyl and has led to a demand for improved standards of radiation protection. One method of improving protection is to establish radioactivity monitoring schemes in the water industry. Now that it is clear that Water Authorities should take radioactivity into account when carrying out their statutory duty to supply wholesome water, people in the industry are expressing a greater interest in radioactivity monitoring (Castle 1988).

Monitoring schemes should be capable of detecting both man-made and natural radionuclides. Natural radionuclides include radon, radium and uranium, which are frequently found in low concentrations in ground waters. The radiation dose attributable to presence of these radionuclides in water is frequently far more significant than that due to man-made radioactivity and may go undetected if incorrect monitoring equipment is used.

Radionuclide standards

This has been discussed by Castle (1988). Radionuclide standards may be derived from an approach laid down by the International Commission on Radiological Protection (ICRP 1985–86). This is a non-government organization which has published recommendations for the protection of the population against ionizing radiations for over half a century.

There are two central requirements laid down by the ICRP which relate directly to the derivation of standards in drinking water:

1 All exposures should be kept as low as reasonably achievable, economic and social factors being taken into account. This is known as the ALARA principle.
2 The radiation dose to individuals, more correctly referred to as the dose equivalent, shall not exceed the limits for appropriate circumstances recommended by the Commission.

The limits for the general public referred to are expressed in terms of the unit of radiation known as the sievert (Sv) or thousandths of a sievert (mSv). The principal limit to members of the public is $1\,\text{mSv year}^{-1}$ although $5\,\text{mSv}$

year^{-1} is permissible in some years provided that the average annual exposure does not exceed 1 mSv year^{-1}. The levels are continually under review and in the UK the National Radiological Protection Board (NRPB) have recently stated that the effective dose equivalent to the public should not exceed 0.5 mSv year^{-1} from any one source. The level of 5 mSv year^{-1} is frequently taken as the level above which some form of emergency action should be considered to limit the dose to the public. In both normal and accidental conditions the ALARA principle must be applied to keep doses to a minimum.

Erring on the side of extreme safety the WHO have provided values in becquerels (Bq) for gross alpha and gross beta activity of 0.1 and 1 Bq l^{-1} respectively. These levels were derived by considering the most radiologically hazardous alpha and beta emitters likely to be encountered in water (radium 226 and strontium 90 respectively). The values are therefore very conservative below which water can be considered potable without any further radiological examination and correspond to a maximum dose commitment of about 0.05 mSv year^{-1} (Newstead 1988).

The NRPB derived emergency reference levels, shown in Table 8.1 for drinking water, give an indication of the concentration levels and times at which substitution of fresh water supplies should be considered following a major nuclear accident in order to avoid exceeding the annual dose of 5 mSv (Newstead 1988).

Table 8.1. *Derived initial concentrations in drinking water (Bq/l) for substituting fresh supplies at various times*

Radionuclides	Limiting water substitution times (days)			
	2	*7*	*14*	*100*
^{89}Sr	2.3×10^5	6.9×10^4	3.6×10^4	8.5×10^3
^{90}Sr	2.8×10^4	8.1×10^3	4.1×10^3	5.7×10^2
^{103}Ru	8.5×10^5	2.6×10^5	1.4×10^5	3.6×10^4
^{106}Ru	7.3×10^4	2.1×10^4	1.0×10^4	1.6×10^3
^{131}I	1.1×10^4	3.7×10^3	2.4×10^3	1.7×10^3
^{133}I	$8.3 + 10^4$	6.7×10^4	6.6×10^4	6.6×10^4
^{134}Cs	4.3×10^4	1.2×10^4	6.2×10^3	9.1×10^2
^{137}Cs	5.1×10^4	1.5×10^4	7.3×10^3	1.0×10^3

The above values apply to water samples at the point of consumption. In all cases, the most restrictive age group is 1-year-old infants. The levels represent an effective whole body dose equivalent of 5 mSv, but do not allow for the additivity of radiation dose from sources other than water. The Table is drawn from NRPB publication and forms the basis of the derived emergency reference levels (DERLs).

In the aftermath of the Chernobyl accident it was recognized that there was a need for unified approach in the EEC to controlling the levels of radioactivity in water and food (Commission for European Communities 1987). Therefore, new levels have been derived, based on an annual dose limit of 5 mSv.

Sources of radioactivity in water

Fallout and nuclear discharges

While there have been several major accidents involving nuclear power stations, the Chernobyl accident was undoubtedly the most dramatic and has provided a great deal of information about the likely effect of any future major reactor accident. It is clear that under the right meteorological conditions, radioactive fallout can be expected to travel thousands of kilometres, crossing international boundaries.

After the Chernobyl accident many measurements were made of the levels of radioactivity in water resources in western Europe and the USSR. In western Europe the radioactive burden of surface water remained low enough to leave drinking water supplies unaffected, although it was necessary to place some restrictions on drinking rainwater. The levels of radioactivity in surface waters remained low and the effect of water treatment was to remove a proportion of the small amount of radioactivity present in the raw water and to concentrate it in the waste sludge. The resulting waste sludges were comparatively radioactive. A similar effect was observed in sewage sludges, although the levels of activity in the sludge were at least an order of magnitude lower. In the USSR and closer to the site of the reactor accident, water supplies were affected and it was necessary to carry out a series of major remedial measures to limit the contamination.

Natural radioactivity

To put man-made radiation into perspective it is important to realize that in normal circumstances it contributes less than 2% of the average person's radiation dose (Castle 1988).

The natural radionuclides of most concern to drinking-water supplies are radium, uranium and radon. Many naturally occurring radionuclides such as uranium 238, uranium 235, thorium 232 and potassium 40 are very slow to decay and represent the remains of the radioactivity produced before or when the earth was formed. Most of the other natural radioisotopes consist of three series of radioisotopes supported by the decay of uranium 238, uranium 235 and thorium 232. More than one-half the radionuclides in these three series are alpha emitters with decay energies of 4 to 8.9 MeV. Other naturally occurring radionuclides such as tritium result from the continuous bombardment of the earth's atmosphere by cosmic rays.

The concentration of a particular radionuclide in the water supply depends upon the concentration of the parent radionuclide in the surrounding rock, the isotope's half-life and its geological setting. The geological setting affects the solubility of the radionuclide, the rate at which the radionuclide escapes from the source rock and the adsorption of the radionuclide by surrounding materials. Ground waters are likely to contain the highest concentrations of natural radionuclides.

Uranium   An excess of uranium 234 occurs in natural waters (higher than that produced by the decay of uranium 238) due to the alpha recoil propelling the uranium 234 into solution. Analytical reports often quote the ratio in water of uranium 234/uranium 238 and this ranges from 1 to about 30. In areas of granite, around Devon, water supplies have been found to contain a little under $2 \, \text{mg} \, l^{-1}$ of uranium.

The decay of uranium 238 gives rise to series of radioisotopes and these include the isotopes of radium 226 and radon 222. Other isotopes of radon and radium are formed in two other decay series but usually the predominant radioisotope in water is radon 222. Typically, the levels of radon 222 in water are three orders of magnitude higher than those of radium 226.

Radon 222   Radon 222 is formed in rock strata by the alpha decay of radium 226. The radon gas diffuses through the imperfections in the rock's structure and into the surrounding air or water. The physical condition of the rock matrix appears to play a greater role than the concentration of the parent radionuclide. It has been noted that weathered igneous rocks are most likely to give rise to high radon concentrations. Radon is very soluble in water and concentrations as high as $10\,000 \, \text{Bq} \, l^{-1}$ have been recorded in some ground waters in the USA.

If radon is present in the water together with radium then the radon is said to be supported. This is because the decay of the long-lived radium produces a continuous supply of new radon. For the first few hours after sealing a sample of water containing radon 222 there is an increase in alpha activity. This results from the in-growth of the short-lived daughter isotopes such as polonium 218 (half life 3.05 min) from the decay of radon (half life 3.82 days).

Radon in water presents the dual pathway of exposure of individuals by ingestion from the direct consumption of water and by inhalation exposure when the radon gas emanates from water. The annual dose resulting from a radon exposure of about $40 \, \text{Bq} \, l^{-1}$ has been estimated to be about 1 mSv (Cross 1984) to the stomach and about $20 \, \mu\text{Sv}$ to the whole body.

The primary cause of elevated concentrations of airborne radon in most homes is the entry of soil gas through the foundations and walls, although in some areas groundwater can be a major source. The effective dose from drinking water has been estimated to be between 1 and 12% of the total annual effective dose from all indoor radon (Cothern 1986). In the USA it has been estimated that radon present in water accounts for ten times as much morbidity as radium and 20 times as much as uranium (Pritchard 1987). It is therefore important to assess the levels of radon in the groundwater sources (and confined spaces in the associated treatment plants) and if necessary to provide treatment to remove radon from the water supply.

Measurement of alpha, beta and gamma radiation

The water industry is concerned with alpha particles, beta particles and gamma rays.

Unfortunately, there is no single probe available which can be placed in a sample of water and which will produce a reading of the dose that a person would receive if he were to drink the water. This is because the hazard posed by ionizing radiation varies both with the type of radiation and with the energy at which the radiation is emitted. For example, it is generally true to say that alpha-particle emitters, when taken orally, are a greater radiation hazard than beta or gamma emitters, although the chemical nature of the substance in which the radioactivity is taken will influence its absorption by the body and hence the radiation hazard. The situation is further complicated because many radiation detectors are designed to measure one type of radiation and often the sensitivity of the detector varies across the energy range of the radiation.

8.1 Instrumentation for alpha, beta and gamma measurements

A wide range of instrumentation is available (Appendix) for measuring alpha, beta and gamma radiation by the gross or more detailed methods discussed above. Individual spectrometers for each type of radiation or combined alpha–beta–gamma counting systems are available. Discussion here will be limited to the Canberra models 2401 and 2401 F (manual) and models 2400 and 2400F (automatic) low level alpha–beta–gamma counting system, as this has been particularly recommended for use in the water industry. This system provides a highly efficient sealed proportional sample counter along with a sealed anti-coincidence guard detector to provide extremely low background sample counting. Combined with a microprocessor-based system controller, the 2400 provides automatic sample changing for up to 100 samples, automatic background, crosstalk, concentration, chi-square, data average and user-entered calculations, all in addition to the operator-definable counting sequences.

This system is ideally suited for measuring nuclides such as ^{14}C, ^{228}Ra, ^{90}Sr, ^{99}Tc, ^{226}Ra, ^{210}Po, ^{230}Th, etc. For special counting requirements, ultra-thin window-flow sample detectors are available, providing the user with very efficient detection of environmental alpha emitters. The system performs simultaneous alpha, beta and gamma counting.

The exclusive features and operating characteristics of the automatic model 2400 alpha–beta–gamma counting system are also available in the efficient, single-sample, manual model 2401. The system employs two main counting channels for alpha and beta events. Energy discrimination is accomplished through a programmable single-channel analyser contained in the controller. Cosmic and other background events are excluded from either sample channel through anti-coincidence circuitry connected to the guard detector. Live-time correction is provided during all guard channel events. An integral detector assembly is included in each system.

Gross radioactivity measurements are a simple and fairly inexpensive method of assessing the level of radioactivity in water. If the results are below the World Health Organisation (WHO) levels of gross alpha–gross beta activity discussed below, generally the water sample can be classified as fit for consumption without further analysis.

It is important to remember that radioactive isotopes such as tritium and radon will not be detected by these methods, and should the level of radioactivity detected exceed the WHO limits then the samples will have to be analysed for these.

Gross measurements

Beta ray spectrometers

The gross beta activity of water samples can be measured by the use of a Geiger counter or proportional counter and scaler (*Parr Manual* 1974; Nadkarni 1981). The sample is evaporated to low volume and then to dryness on a tray or planchette which is presented to the counter. Results have to be corrected for background and efficiency of detection.

The gross alpha activity of a sample can be measured in a similar manner using a zinc sulphide scintillation counter or proportional counter, but the problems of the self-absorption of alpha particles are considerably greater than those encountered with beta particles and this must be allowed for in the calculations of the analytical figure.

Gamma ray spectrometers

Gamma-ray spectrometers are used to identify and quantify the level of individual gamma-emitting radioisotopes in a sample. Basically there are two types in common use which differ in the nature of the detector. The type based on the sodium iodide type of detector is the least expensive but suffers from the considerable drawback that the spectrum produced is very poorly resolved and this limits the ability of the technique to identify gamma emitters. The type based on the hyper-pure germanium detector has excellent resolution, but has to be kept at liquid nitrogen temperatures and this type of detector is best suited to the type of analysis required in the water industry.

8.2 Instrumentation for radon monitoring

Liquid scintillation counters are used for a variety of radioactivity measurements, of which the most important for the water industry are the measurement of tritium and radon.

Various methods of assessing radon concentrations in water have been developed. In one such method the radon in a sample of water is partitioned into toluene, which is then mixed with a scintillant and counted in a liquid scintillation counter. This method has the advantage of being relatively simple in principle, although there are many practical difficulties. One of the major difficulties is taking the sample. Radon is readily dissipated from water as it emanates from the tap and therefore samples must be taken in sealable bottles in such a way that turbulence while filling the bottle is minimized.

Packard supply an instrument which operates on these principles – the Pico Rod System (see Appendix).

The Packard Tricarb 1900 CA liquid scintillation analyser used in the Picorad system is computer controlled and designed to operate unattended. It features a built-in IBM-compatible computer with the optional band disk and has a monitor on a movable arm. The apparatus is capable of handling 200 samples per day. The Packard Tricarb 1500 or the top-of-the-range Tricarb 2200 CA analysers can also be used in this application. The Tricarb 1900 CA and 2200 CA analysers provide a live spectrum of the sample being counted. After samples have been counted the software automatically calculates results, processes reports and maintains a record of results on disk. A Pico rad quality-assurance program is in operation for users of these instruments.

8.3 Instrumentation for tritium measurement

NR Technology Ltd supply the model RGM1/1 tritium monitor. It measures the activity of alpha- and beta-emitting radioactive gases in air in the presence of gamma-emitting radiation background.

Portable multi-channel analysis systems

Canberra supply the Series 10 plus battery-operated portable multi-channel analyser system which it is claimed is suitable for environmental surveys. There is a choice of an Na I(Tl) detector or a portable germanium detector for field measurements of gamma-emitting species in on-site water samples. The system weighs 5.3 kg, including a high-resolution LCD display for both manual and autoranging linear and log display. Cassette data storage is included. By adding a 1184 PC interface and a 1 MB PC the series 10 becomes the portable front end of a powerful Saturn gamma assay system.

Table 8.2. *Removal efficiencies of radionuclides by water-treatment processes*

Water treatment	^{51}Cr	^{32}P	^{90}Sr	^{91}Y	^{106}Ru	^{131}I	^{137}Cs	^{144}Ce	^{222}Rn	^{226}Ra	^{239}Pu	Particulate-associated radioactive compounds
Aeration									90–100			
Chemical coagulation, settling and/or filtration	10–98	68–99	0–70	80–91	77–96	0–44	0–6	80–94				85–90
Slow sand filtration		80–99	0–5			50–99	50	99			90	
Softening with lime soda		99	20–80	90	50	0–10	0–20		85			
Ion exchange			98.5	75–99		0	99		95			
Reverse osmosis			99						92–95			

Values are only estimates based upon information available.

8.4 Applications of radioactivity measurements

Evidence suggests that traditional water-treatment processes are effective in removing a significant proportion of radionuclides from raw water supplies although the type of treatment, chemical nature of the water and the nature of the radionuclides all affect removal efficiency.

Estimates of the removal efficiency of radionuclides by water-treatment processes are shown in Table 8.2. Generally speaking radionuclides such as those of ruthenium and caesium, which are quickly absorbed onto fine particulate matter in the water, seem to be readily removed by filtration, coagulation and settling processes while the more soluble radionuclides such as those of strontium and iodine pose a greater problem, and ion-exchange resins or reverse osmosis may be required.

The Chernobyl accident provided information about the removal efficiencies under 'real-life' conditions. At Arnfield treatment plant, east of Manchester in the UK, removal efficiencies of 80% gross beta activity were recorded, and in the Netherlands removal efficiencies measured at six surface water treatment plants were found to vary from 50 to 84%.

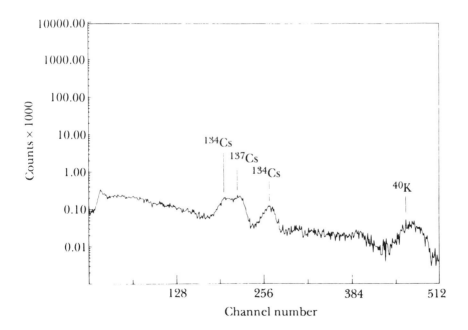

Figure 8.1 *Gamma spectrography indicating fall-out products in environmental sample*

Gamma spectrometry is undoubtedly the technique to use when it is required to identify and determine individual radioactive isotopes in water samples, sludges or sediments. The multi-channel analyser trace in Figure 8.1 shows the presence of ^{137}Cs, ^{134}Cs and naturally occurring ^{40}K.

9 On-site measuring instruments

This chapter discusses available portable instrumentation that can be carried around by inspectors to carry out measurements on rivers, reservoirs, effluents, etc.

9.1 Rapid-test kits

Merck supply a range of test kits covering many different determinands (Table 9.1). Test strips and test kits, both visual and photometric are available and all are suitable for use in the field.

Merckoquant test strips

These consist of a reactive test zone firmly bonded to a plastic backing. The test zone is impregnated with reagents, buffers and other substances. The strips are used for rapid exploratory testing of substance concentrations as low as $1 \, mg \, l^{-1}$. To test a sample the test zone is immersed in the water sample for 1–2 s and matched against a colour scale.

Aquamerck test kits

These consist of titrimetric and colorimetric test kits supplied in boxes or blister packs. The simplest titration tests contain a dropping bottle or precision dropping pipette. The number of drops of reagent required to change the colour of the indicator is a measure of the concentration of the substance being tested for. Aquamerck colorimetric tests incorporate a waterproof scale with precise directions, or alternatively a testing vessel consisting of a 10 mm cell with coloured reference blocks to the right and left of it so that a colour comparison can be made.

Aquaquant test kits

The Aquaquant system for rapid water analysis fully meets the stringent requirements relating to detection sensitivities, ease of use and economy. Each Aquaquant test kit consists of a plastic case containing a sliding colour

Table 9.1. Test kits supplied by Meeck

Determination	Measuring range or graduation mg/l (p.p.m.)	Merckoquant® test strips	Aquamerck® titrimetric test with dropping bottle or titrating pipette	Aquamerck® colorimetric test in a blister pack	Aquamerck® colorimetric test with colour scale	Aquamerck® colorimetric test with testing vessel	Exchange pack	Aquaquant® test with sliding colour comparator	Microquant® test with comparator disc	Spectroquant® test for rapid photometric determinations
Acid binding capacity (ABC)	0.25–20 mmol		11147							
Acidity	0.1–40 mmol/0.1		11108							
Alkalinity	0.1–40 mmol/0.1		11109							
Aluminium	10–250	10015								
	0.07–0.8							14413	14822	
	0.1–8.0									14825
	0.06–1.4						14824			
Ammonium (Nessler method) (indophenol blue method)	10–400	10024								
	0.05–0.08									
	0.5–10			14657				14400		
	0.2–5				8024					14752
	0.025–0.4							14428		
	0.2–8.0							14423	14750	
	0.2–8.0									
	0.03–3.0									
Arsenic	0.1–3	10026								
Ascorbic acid	50–2000	10023								
Boron	0.03–0.8								14837	14839
	0.008–2.0									
Calcium	25–250	10034								
	2–800/2		11110							
	3–40							14443	14813	
	10–200									14815
	5–300									
Carbonate hardness	1–100°d/l		14653							
	1–100°d/l		11103							
	0.2–80°d/0.2		8048				8041			

	Range								
Chloride	25–2500/25 2–800/2 5–300 3–300 0.4–40		11132 11106				14401	14753	14755
Chlorine	4–120 0.01–0.3 0.25–15 0.05–3 0.1–2.0 0.05–5	10043					14434	14826	14828
in swimming pools				14670				14801	14803
Chlorine and pH (DPD method)	Chlorine 0.1–1 pH 6.5–8 Chlorine 0.1–1 pH 6.8–7				11160	11135	11157 11143		
(o-tolidine method)	Chlorine 0.1–1 pH 6.8–7					11134	11133 11143		
Chromate	3–100 0.005–0.1 0.1–1.6 0.1–10 0.025–2.5	10012					14402 14441	14756	14758
Cobalt	10–1000	10002							
COD	10–150 100–1500						14421		
Colour (water colouration)	5–150 Hazen								
Copper	10–300 0.15–1.6 0.05–0.5	10003		14651					
Nitrite	0.1–10 0.03–3						14414	14774	14776
Oxygen	1–12 0.1–30/0.1 0.1–30/0.1		11107 11149	14662*		11152 11152			
Oxidizing agents (potassium iodide-starch paper)	0.05–0.19°e	9512							14540* 14541*
Permanent hardness				11142					
Peroxide	1–100 aqueous solution 1.5–80 organic solution	10011							

Table 9.1. (*Continued*)

Determination	Measuring range or graduation mg/l (p.p.m.)	Merckoquant® test strips	Aquamerck® titrimetric test with dropping bottle or titrating pipette	Aquamerck® colorimetric test in a blister pack	Aquamerck® colorimetric test with colour scale	Aquamerck® colorimetric test with testing vessel	Exchange pack	Aquaquant® test with sliding colour comparator	Microquant® test with comparator disc	Spectroquant® test for rapid photometric determinations
Perex test	10–500				16206					
pH	4.5–9.0/0.5				8027	8038	8043			
	4.5–9.0/0.5						8043			
in fresh water	5.0–9.0/0.5			14655						
in sea water	7.1–8.9/0.3			14656						
in swimming-pool water	6.5–8.2			14669						
in swimming-pool water	6.8–7.8									
pH universal indicator liquid	4–10/0.5				9175		11143			
	4–10/0.5				9175					
pH indicator liquid	0–5/0.5				9177					
	9–13/1.0				9176					
pH indicator liquid	4.5–10/0.5							14425		
	6.4–8.6/0.2							14430		
	5.2–7.4/0.2							14436		
Phosphate (vanadate–molybdate method)	2.2–29 (P_2O_5) 5.67 (Na_3PO_4)									
	1–40					8016	11125	14449	14840	14842
	1.5–100						11125			
	0.25–25									
(ammonium heptamolybdate method)	0.1–2.0			14661		11138	8046	14409	14786	14788
	1–10 (P_2O_5)						8046			
	0.01–0.16									
	0.1–3.0									
	0.024–2.4									
Phosphate silicate	1–10 (P_2O_5) 0.3–3 (SiO_2)					11119	11123			
							8022			
Potassium	300–2000	10042								
Silicon	0.01–0.25							14410	14792	14794
	0.3–10									
	0.03–8.0									

Determinand	Range					
Silver (fixing bath test)	0.5–10 g/l	10008				
Sulphate	200–1600	10019		14411	14789	14791
	25–300					
	25–300					
	10–600					
Sulphide (lead acetate paper)		9511				
Sulphite	10–500	10013	11148			
	2.5–200/2.5					
Tartaric acid	0.5–10 g/l	10021				
Thioglycollate detection paper		9576				
Tin	10–200	10028				
Total hardness	4–29°e	10025				
	4–29°e	10032				
	4–29°e	10029				
	6–31°e	10046				
	6–31°e	10047				
	1–100°e/1.25		14652			
	1–100°e/1.25		11104			
	1–100°e/1.25		8011			
	1–100°e/1.25		11111			
	0.2–80°e/0.2		8039			
	0.1–15°e/0.1		8047			
Titrant solutions for 8011, 11111, 8039			8033	11141		
			8040			
Indicator tablets for 8011				11140		
Indicator solution for 8039, 8047, 11111				11122		
Water colouration see 'Colour'						
Zinc	10–250	10038	14782	14412	14780	
	0.1–5.0					
	0.1–5.0					

comparator, which is of a moulded construction designed to hold test solutions, reagents and accessories as a test is being performed. Short-tube and long-tube test kits are available. The different path lengths allow for lower and higher sensitivities.

Microquant test kits

There is often a need for a rapid colorimetric test in a medium sensitivity range roughly equivalent to that covered by the Aquamerck and Aquaquant short tube tests, but which will also permit coloured or turbid samples to be measured. The Microquant tests, which work with transmitted light, are eminently suitable for this type of analysis.

Spectroquant analysis system

This comprises an SQ 115 digital photometer, a TR 205 Thermoreactor for elevated temperature test, 100 and 148°C, and the Spectroquant test kits. The SQ 115 photometer measures absorbance and concentration in the range 370–1000 nm. The photometer is conveniently calibrated with calibration cells, which incorporate transparent coloured windows to simulate standard solutions of precisely defined concentrations. The inconvenience of calibrating against standard solutions and the unreliability of using calibration factors are thus avoided.

Merck also supply compact laboratories consisting of packs of reagent bottles and accessories. These include the determination of ammonium, carbonate, hardness, nitrite, nitrate, pH and dissolved oxygen.

Palintest rapid test kits

These are available in a visual test form (tablet count, colour match and turbidity tests) or as a spectrophotometric version using the Palintest Photometer 5000 (Table 9.2).

Palintest also supply swimming pool and spa test kits for the determination of free chlorine $(0-3 \, \text{mg} \, l^{-1})$, total bromine $(0-8 \, \text{mg} \, l^{-1})$, pH $(6.8-8.4)$ total alkalinity and calcium hardness $(0-1000 \, \text{mg} \, l^{-1})$, cyanuric acid $(0-200 \, \text{mg} \, l^{-1})$, chloride $(0-5000 \, \text{mg} \, l^{-1})$, copper $(0-0.9 \, \text{mg} \, l^{-1}$ or $0-4 \, \text{mg} \, l^{-1})$, sulphate $(0-200 \, \text{mg} \, l^{-1})$, iron $(0-1 \, \text{mg} \, l^{-1})$ and total dissolved solids $(0-1999 \, \text{mg} \, l^{-1})$. Comparator 2000 and photometer 5000 versions are available.

De Lange cuvette and pipette tests

These are all based on the use of portable spectrophotometers, the LASA Aqua filter photometer for small numbers of water analyses, the LPIW filter photometer for water and sewage analysis and the top-of-the-range Cadas 100 with print-out of results for water and sewage analysis (Table 9.3). The latter instrument covers the spectral range 200–900 nm amongst many others. Chemical oxygen demand, ozone and formaldehyde determinations can be carried out with this equipment.

9.2 Probe or dipstick measurements of pH, electrical conductivity, total dissolved solids, temperature and turbidity

These are discussed under one heading as several instruments are available which measure more than one of these parameters.

Multi-parameter instrument

Horiba supply the portable battery-operated model U-7 series water-quality checkers which measure pH, dissolved oxygen, electrical conductivity, temperature and turbidity.

Specifications for the Horiba instrument are given in Table 9.4. In addition to this instrument Horiba also supply single-parameter instruments, the U-7-pH, the U-7-Do and the U-7 conductivity.

CP Instrument Company also supply a multi-parameter instrument, the JN3405-00 portable electrochemistry analyser, which measures pH, mV, temperature, conductivity (five ranges) and dissolved oxygen (Table 9.5).

PHOX supply portable 4–6 channel water-quality monitors covering any 4–6 of the following parameters:

- dissolved oxygen
- temperature
- pH or redox
- conductivity and total dissolved solids
- water flow, volume or level
- suspended solids or turbidity
- nitrate
- ammonia
- organic pollutants

PHOX also supply mobile monitoring systems based on a towable trailer with a selection of monitoring equipment tailored to meet user's requirements.

Single-parameter instruments

A range of available instrumentation is tabulated in Table 9.6.

Table 9.2. *Test kits available from Palintest*

Order Code	Test	Range	Test method
PS 400	Aluminium	$0–0.5$ mg l^{-1}	CM
PS 405	Ammonia	$0–0.8$ mg l^{-1} N	CM
		$0–1.0$ mg l^{-1} NH$_4$	CM
PS 410	Chlorine (DPD)	$0.2–1.0$ mg l^{-1}	CM
		$1.5–8.0$ mg l^{-1}	CM
PS 411	Chlorine (KI)	$10–160$ mg l^{-1}	CM
PS 415	Copper	$0–0.8$ mg l^{-1}	CM
		$1.0–4.0$ mg l^{-1}	CM
PS 425	Hydrazine	$0–1.0$ mg l^{-1}	CM
PS 430	Hydrogen peroxide	$0.2–100$ mg l^{-1}	CM
PS 435	Iron	$0–10$ mg l^{-1}	CM
PS 440	Nitrate	$0.2–1.0$ mg l^{-1} N	CM
		$0.88–4.4$ mg l^{-1} NO$_3$	CM
PS 445	Nitrite	$0.05–0.5$ mg l^{-1} N	CM
		$0.15–1.6$ mg l^{-1} NO$_2$	CM
PS 450	Phosphate	$0–80$ mg l^{-1} PO$_4$	CM
PS 455A	pH	$6.0–9.2$	CM
PS 455B	pH	$3.0–12.0$	CM
PS 465	Zinc	$0–4.0$ mg l^{-1}	CM
CS 112	Total alkalinity	$0–500$ mg l^{-1}	TC
CS 119	Calcium hardness	$0–500$ mg l^{-1}	TC
CS 113	Chloride	$0–1000$ mg l^{-1}	TC
CS 126	Chlorocol (chlorine)	$0–50$ mg l^{-1}	CM
CS 133	Cleaning acid strength	$0–10\%$	TC
CS 127	Cyanuric acid	$0–200$ mg l^{-1}	TU
CS 118	Hardness VLR	$0–5$ mg l^{-1}	TC
CS 117	Hardness LR	$0–50$ mg l^{-1}	TC
CS 116	Hardness	$0–500$ mg l^{-1}	TC
CS 130	Hardness Yes/No	4.8 or 20 mg l^{-1}	YN
CS 120	Nitrite	$0–1500$ mg l^{-1}	TC
CS 138	Organo-phosphonate	$0–20$ mg l^{-1}	TC
CS 129	Universal pH	$4–11$	CM
CS 131	Phosphate limit	$30/70$ mg l^{-1}	CM

Table 9.2. (*Continued*)

Order Code	Test	Range	Test method
CS 139	Quatest (QAC)	0–200 mg l^{-1}	TC
CS 132	Sulphate	0–3000 mg l^{-1}	TU
CS 123	Sulphite LR	0–50 mg l^{-1}	TC
CS 124	Sulphite HR	0–500 mg l^{-1}	TC
PM 166	Aluminium	0–0.5 mg l^{-1}	
PM 152	Ammonia	0–1.0 mg l^{-1} N	
PM 060	Bromine	0–6.0 mg l^{-1}	
PM 011	Chlorine (free, combined and total)	0–5.0 mg l^{-1}	
PM 162	Chlorine HR (total)	0–250 mg l^{-1}	
PM 084	Copper LR	0–1.0 mg l^{-1}	
PM 083	Copper HR	0–4.0 mg l^{-1}	
PM 086	Copper (total inc chelated)	0–4.0 mg l^{-1}	
PM 087	Cyanuric acid	0–200 mg l^{-1}	
PM 179	Fluoride	0–1.5 mg l^{-1}	
PM 104	Hydrogen peroxide LR	0–2.0 mg l^{-1}	
PM 105	Hydrogen peroxide HR	0–100 mg l^{-1}	
PM 155	Iron LR	0–1.0 mg l^{-1}	
PM 156	Iron HR	0–10 mg l^{-1}	
PM 173	Manganese	0–0.1 mg l^{-1}	
PM 175	Molybdate HR	0–100 mg l^{-1} MoO_4	
PM 163	Nitrate	0–1.0 mg l^{-1} N	
PM 109	Nitrite	0–0.5 mg l^{-1} N	
PM 056	Ozone	0–2.0 mg l^{-1}	
PM 130	pH	6.8–8.4	
PM 177	Phosphate LR	0–4.0 mg l^{-1}	
PM 114	Phosphate	0–100 mg l^{-1}	
PM 154	Sulphate	0–200 mg l^{-1}	
PM 168	Sulphide	0–0.5 mg l^{-1}	
PM 148	Zinc	0–4.0 mg l^{-1}	

Table 9.3. *De Lange test kits*

Cuvette tests

	Measuring range	Method	LASA Aqua*	LP1W*	CADAS 100**	Reaction temperature
LCK 303 ammonium	$1–40$ mg l^{-1}	Indophenol blue in analogy to DIN 38 406 E 5	×	×	×	Ambient
LCK 304 ammonium	$0.01–2.0$ mg l^{-1}	Indophenol blue in analogy to DIN 38 406 E 5	×	×	×	Ambient
LCK 308 cadmium	$0.01–0.3$ mg l^{-1}	Cadion	×	×	×	Ambient
LCK 208 calcium	$0.5–10.0$ mg l^{-1}	GBHA	–	×	–	Ambient
LCK 310 chlorine	$0.05–2.0$ mg l^{-1}	DPD in analogy to DIN 38 406 G 4	×	×	×	Ambient
LCK 311 chloride	$1–70$ mg l^{-1}	Iron-III-thiocyanate	×	×	×	Ambient
LCK 313 chromium, total and hexavalent	$0.05–1.0$ mg l^{-1} $0.01–0.25$ mg l^{-1}	Diphenylcarbazide in analogy to DEVE 10	×	× ×	× ×	Ambient and 100°C
LCK 114 COD	$150–1000$ mg l^{-1}	Chromosulphuric oxidation	×	×	×	148°C
LCK 314 COD	$15–150$ mg l^{-1}	Chromosulphuric oxidation	×	×	×	148°C
LCK 315 cyanide	$0.01–0.5$ mg l^{-1}	Barbituric pyridine in analogy to DIN 38 405 D 13	×	×	×	Ambient
LCK 321 iron	$0.2–10$ mg l^{-1} $0.01–0.25$ mg l^{-1}	1,10-Phenanthroline in analogy to DIN 38 406 El	×	× ×	× ×	Ambient
LCK 323 fluoride	$0.1–1.5$ mg l^{-1}	Spadns	×	×	×	Ambient
LCK 325 formaldehyde	$0.5–10$ mg l^{-1} $0.01–2$ mg l^{-1}	Acetylacetone in analogy to DIN 52 368	×	× ×	× ×	40°C
LCK 327 water hardness calcium and magnesium	$1–18°$dH $5–100$ mg l^{-1} Ca $5–50$ mg l^{-1} Mg	o-Cresolphthalein Complexon	× × ×	× × ×	× × ×	Ambient Ambient Ambient
LCK 329 copper	$0.1–4.0$ mg l^{-1} $0.01–0.1$ mg l^{-1}	Bathocuproinedisulphonic acid	×	× ×	× ×	Ambient

Product	Method	Measuring range				Temperature
LCK 337 nickel	Dimethylglyoxime in analogy to DEVE 11	0.1–5.0 mg l⁻¹	×	×	×	Ambient
		0.01–0.1 mg l⁻¹	×	×	×	
LCK 339 nitrate	2,6-Dimethylphenol in analogy to DIN 38 405 D9	3–80 mg l⁻¹	×	×	×	Ambient
LCK 341 nitrite	Sulphanilic acid/naphtylamine in analogy to DIN 38 405 D 10	0.1–3.0 mg l⁻¹	×	×	×	Ambient
		0.01–0.1 mg l⁻¹	×	×	×	
LCK 343 ozone	DPD	0.05–2.0 mg l⁻¹	×	×	×	Ambient
LCK 049 o-phosphate	Vanadate molbydate (VMR)	5–100 mg l⁻¹	×	×	×	Ambient
LCK 349 total and o-phosphate	Phosphomolybdic blue in analogy to DIN 38 405 D 10	0.1–5.0 mg l⁻¹	×	×	×	Ambient and 100°C
		0.01–0.5 mg l⁻¹	×	×	×	
LCK 355 silver	–	5–3000 mg l⁻¹	×	×	×	Ambient
LCK 153 sulphate	Barium sulphate	10–150 mg l⁻¹	×	×	×	Ambient
Pipette tests						
LCW 001 aluminium	Eriochromcyanin in analogy to DEVE 9	0.05–1.0 mg l⁻¹	–	×	×	Ambient
LCW 006 lead	Dithizone	0.01–1.0 mg l⁻¹	–	×	×	Ambient
LCW 017 detergents anion-active	Methylene blue	0.01–2.0 mg l⁻¹	–	×	×	Ambient
LCW 025 hydrazine	4-dimethyl-aminobenzaldehyde	0.01–2.0 mg l⁻¹	–	×	×	Ambient
LCW 028 silicic acid	Molybdenum blue	0.1–0.5 mg l⁻¹	–	×	×	Ambient
LCW 032 manganese	Formaldoxime in analogy to DIN 38 406 E 2	0.02–5.0 mg l⁻¹	–	×	×	Ambient
LCW 053 sulphide	Dimethyl-p-phenylenediamine in analogy DEV 07	0.05–1.0 mg l⁻¹	–	×	×	Ambient
LCW 054 sulphite	Iodide/iodate	0.1–5.0 mg l⁻¹	–	×	×	Ambient
LCW 060 zinc	4-(2-pyridylazo)-resorcinol	0.02–10 mg l⁻¹	×	×	×	Ambient

* digital, ** wavelength range 200–900 μm.

Table 9.4. *Specifications of Horiba H-7 series multi-parameter instrument*

	pH	Temperature	DO	Conductivity	Turbidity
Principle	Glass electrode	Thermistor	Membrane-galvanic cell	Four-electrode sensor	Ratio turbidometer
Range	pH 0–14	0–40°C	0–20 p.p.m.	$0 \sim 50$ mS cm^{-1} (STD) $0 \sim 10$ mS cm^{-1} (option) $0 \sim 1$ mS cm^{-1} (option)	$0 \sim 400$ p.p.m.
Repeatability	± 0.1 pH ± 1 digit	± 0.5°C ± 1 digit	± 1.0 p.p.m. ± 1 digit	± 2.5 mS cm^{-1} ± 1 digit	± 20 p.p.m. ± 1 digit
Resolution	0.01 pH	0.1°C	0.1 p.p.m.	0.1 mS (STD)	1 p.p.m.
Temperature compensation	Automatic, 0–40°C	–	Automatic, 0–40°C	–	–
Meter	Light emitting diodes, 3 digits				
Mode selection	Rotary switch				
Power source	Rechargeable Ni–Cd battery, d.c. 7.2 V (with charger), 2 h continuous operation				
Power consumption	Approx. 1.5 W				
Connection cable	2 m, optional length 10 m				
Weight	Instrument: Approx. 700 g, Sensor: Approx. 900 g				

Table 9.5. *Specification of CP Instrument Co JN 3405-00 portable electrochemistry analyser*

Ranges	Resolution	Accuracy
0 to 14.00 pH	0.01 pH	± 0.02 pH
0 to ± 1999 mV	1 mV	± 1 mV
0 to 100°C	0.1°C	± 1°C
0 to 199.9 mS	0.1 mS	$\pm 3\%$ on 200 mS range
0 to 19.99 mS	0.01 mS	$\pm 0.5\% \pm 2$ digits on all other
0 to 1999 µS	1 µS	conductivity ranges
0 to 199.9 µS	0.1 µS	
0 to 19.99 µS	0.01 µS	
0 to 200%	1%	1%
0 to 19.9 mg l^{-1}	0.1 mg l^{-1}	± 0.2 mg l^{-1}

9.3 Portable trace-metal analysers

These are produced by EDT Analytical Ltd (Model PDV 2000) portable trace-metal analyser.

The technique of stripping analysis or voltammetry has been known for many years. It is in fact a form of electroplating on a small scale. Metal is plated from a solution onto an electrode by applying, for a specified period of time, a negative potential to the electrode. During the subsequent stripping stage, the same electrode has an increasingly positive voltage applied to it (a voltage ramp) and the deposited metal is reoxidized or stripped back into solution. The small amount of current generated as each metal oxidizes is measured and correlated with the concentration of metals in the original solution. The technique is ideally suited to the detection and analysis of metals at trace concentrations in water samples.

The instrument uses a glassy carbon working electrode in an electrochemical cell, and is driven by a microprocessor from a 20-switch keypad.

Conventional polarographic analysis requires that oxygen be removed from analyte solutions usually by bubbling pure nitrogen through them. This cumbersome procedure has been eliminated in the PDV 2000 by the use of a special electrolyte, which simultaneously removes oxygen and provides a suitable medium for analysis. Samples and standards are diluted with this electrolyte prior to analysis.

Table 9.6. *Single-parameter instruments*

Supplier	Parameter	Model No.
Palintest	pH stick meter	PT 151
Palintest	Total dissolved solids stick meter	PT 152 (0–1990 mg l^{-1}) PT 153 (0–10 000 mg l^{-1})
Palintest	Temperature stick meter	PT 154 (−15–170°C) PT 156 (−40–99.9°C) PT 157 (−40–999°C)
Bibby	pH, mV temperature stick meter	SMP 1
Bibby	Electrical conductivity stick meter	SMC 1
Bibby	Dissolved oxygen and temperature meter	SMO 1
PHOX or Jenway	pH	Model 21 pH indicator Model 47 battery-operated recorder Model 42E weatherproof indicator Model 42 indicator with analogue display

PHOX or Jenway	Electrical conductivity or total dissolved solids	Model 57 – battery-operated recorder Model 52E – weatherproof indicator
PHOX or Jenway	Dissolved oxygen and temperature	Model 67 – battery-operated recorder Model 68 TF – weatherproof indicator Model 62 – digital LCD indicator
PHOX or Jenway	Flowmeter	Series 80 battery operated
PHOX or Jenway	pH	3050 prove 3061 stick meter
PHOX or Jenway	pH/mV	3070 or 3100 probe micropressor meters
PHOX or Jenway	Electrical conductivity	4060 stick meter 407 probe meter
PHOX or Jenway	Dissolved oxygen/temperature	9060 stick meter 9070 probe meter

The PDV 2000 portable digital voltammeter has seven built-in pre-programmed analysis routines covering all common metals encountered in water analysis including:

Menu 1: Zinc, cadmium, lead
Menu 2: Cadmium, lead, copper
Menu 3: Antimony
Menu 4: Arsenic
Menu 5: Gold
Menu 6: Thallium, bismuth
Menu 7: Mercury
Menu 8: User programmable

Other metals which can be determined by ASV with the PDV 2000 include manganese, silver, indium, selenium, tin.

Sensitivity is in the range $1\,\mu g\,l^{-1}$ zinc, lead, thallium, bismuth, cadmium, and copper to $10\,\mu g\,l^{-1}$ (antimony, arsenic and mercury).

9.4 Oil in water measurement

Horiba supply the OCMA-220 oil-content analyser which is capable of carrying out on-site determinations of oil in water.

The analyser is a self-contained, portable instrument, designed to analyse industrial wastewater for hydrocarbon contamination (oils, greases, fats, waxes).

The instrument provides a digital display of oil-in-water contamination to within one part $mg\,l^{-1}$. Either a $0–5\,mg\,l^{-1}$ or $0–20\,mg\,l^{-1}$ measuring range may be selected. Operator-oriented controls allow the user to sequence each phase of the analysis procedure. Measured amounts of sample water and Freon solvent are mixed automatically and allowed to separate. The solvent extracts all hydrocarbon traces from the water and the oil content is accurately measured by a non-dispersive infrared analyser.

10 On-line process measuring instruments

On-line process analysers are used throughout the water industry in applications such as the monitoring of rainwater, water in distribution systems, potable water and effluent and sewage treatment processes. A wide range of instrumentation is available from various suppliers. Some instruments determine single parameters and some are multi-parameter instruments. It is convenient to discuss these under separate headings. It is of interest to note that the Water Research Centre, Medmanham, UK has set up four test sites in the UK for the evaluation of new technology under operational conditions for the evaluation of instrumentation (including flow, pH, chlorine residual, dissolved oxygen and ammonia meters) and computer-based control systems.

10.1 Single-parameter instrumentation

10.1.1 pH

Available industrial instrumentation is reviewed in Table 10.1 and the Appendix.

In its simplest application, pH measurement of, e.g. treated water provides an indication of its acidity and alkalinity. A permanent record of the pH of a sample and so an output signal is normally provided from the pH amplifier to a recorder. In more sophisticated systems, input may be made to a controller, data logger or computerized control system. In most cases changes of pH signify the need for a change in the quantity of reagent being fed into the sample; very often the pH-measuring equipment can be used as the primary element of a closed-loop automatic control system directly controlling pumps or valves (as shown in the Kent Instrumentation system).

10.1.2 Electrical conductivity

Suppliers of industrial electrical conductivity meters are listed in Table 10.2.

10.1.3 Dissolved oxygen

Horiba supply the model WAXA-100 industrial dissolved oxygen meter (Table 10.3). This meter offers precise determination of the dissolved oxygen

Table 10.1. *Industrial instrumentation for single-parameter measurements of pH*

Parameter	Supplier	Model No.	Description
pH	PHOX Systems Ltd and Jenway Ltd	PHOX 40 PHOX 45 3080	Panel-mounted indicator/controller weather-resistant panel-mounted pH motor with recorder output
		3080 HL	Panel-mounted pH meter with recorder output and Hi–Lo alarms and relay output
		3090	Water-resistant pH meter with recorder output
		3090 HL	Water-resistant pH meter with recorder output and Hi–Lo alarms and relay output
PH and Redox	Ingold AG	524	pH and Redox immersion probes operating up to 100°C for sewage plants
		502	pH and Redox immersion probes operating up to 100°C for sewage plants
		515	pH and Redox immersion probes operating up to 100°C for sewage plants
		521	pH and Redox immersion probes operating up to 100°C for sewage plants for use in severe conditions (heavy pollution, organic components, etc.)
pH and Redox	Kent Industrial Measurements	9140 series	Series pH meters give clear, accurate digital read-out combined with a number of output and control functions. The Series comprises four basic instruments, Models 9141, 9142, 9143 and 9144, offering from indicator only through on/off control to mark–space control.

pH and Redox	Kent Industrial Measurements	9160 series	A versatile panel mounting unit available with 0–14 pH/0–800 mV scales or any 8, 10 or 12 pH span. High and low alarms, manual and automatic temperature compensation and outputs of 0–10 mA or 4–20 mA.
		9170 series	General-purpose wall-mounting unit. 0–14 pH/0–800 mV (0–10 or 2–12 pH optional). Case weatherproof to IP55, high and low alarms, manual/automatic temperature compensation, 0–10 mA or 4–20 mA output.
		9180 series	Versatile top-range instrument featuring a demountable pre-amplifier. Operates on any 2, 5 or 10 pH range in addition to 0–14 pH, and has high and low alarms and an isolated output of 0–1, 0–10, 0–20 or 4–20 mA.
		Intrinsically Safe pH system (BASEEFA approved/ EEx ia IICT4, Tamb = 50°C) Model 9188 certificate No. Ex 832309 Model 9189 certificate No. Ex 832310	The System comprises Model 9189 expanded scale pH transmitter, and Model 9188 power supply unit. Model 9189 is used with a pre-amplifier unit, enabling distances of up to 1 km to be covered between electrodes and transmitter. Switch-selectable ranges of 2, 5 or 10 pH units or 14 pH span are available on this instrument. Model 9188 power supply includes integral signal isolator, with high and low alarm set points.
		System 19 modules for building up multiple systems	A rack-mounted modular system intended for used in multi-parameter or multi-channel applications.
		P96M pH controller	Analogue PID controller with optional balanceless and bumpless auto-manual changeover, or motorized valve control with relay outputs and optional auto-manual. Relay or open-drain control in on/off and time proportioning PID forms.

Table 10.1. (*Continued*)

Parameter	Supplier	Model No.	Description
		2867 electrode systems	Simple ABS dip system using a combination pH or redox electrode. Mounts on a 1 in B.S. flange or wall-mounting bracket. Operates up to 70°C.
		7600 electrode systems	Electrode system in polypropylene and glass-coupled polypropylene. In-line, flow and dip variants are available. pH and redox electrodes may be used with a long life moulded reference electrode. Electrode cleaning systems are available.
		7620 electrode systems	Similar to the 7600 Series but manufactured in polyvinylidene fluoride (PVDF) for enhanced chemical resistance, particularly to chlorine/chloride-bearing samples. Dip and flow systems are available. A range of stainless steel systems is available. Variants for flow (2850), dip (2869) and in-line (2879) mounting are manufactured and a withdrawable steam-sterilizable unit (2891) is available.
		7670 electrode systems	Simple low-cost unit in PVC for in-line mounting of 1180 type plastic combination electrode. Available for ½ inch i.d. or 20 mm o.d. pipe.
pH and Redox	Kent Industrial Measurement	Electrode cleaning system 7610/11	Mechanical cleaning system for 7600 Series systems. A polypropylene brush wipes the pH or redox electrode continuously or at pre-determined intervals.
		7612/13	Ultrasonic cleaning unit for 7601 (flow) or 7604/5 (dip) electrode systems. A specially designed electrode ensures undiminished electrode life. Ultrasonic units are also available for stainless steel electrode systems.

Table 10.2. *Industrial instrumentation for single-parameter measurement of electrical conductivity*

Parameter	Supplier	Model No.	Description
EC	PHOX	PHOX 50 PHOX 55	Panel-mounted EC indicator controller weatherproof transmitter/indicator controller
		4080	Panel-mounted EC meter with recorder
		4080 HL	Panel-mounted EC meter with recorder and Hi–Lo alarm and relay output
		4090	Water-resistant EC meter with recorder
		4090 HL	Water-resistant EC meter with recorder and Hi–Lo alarm and relay output
EL	Horiba	WACA 120	EC measurement on untreated water and sea water $(0-1000\,\mathrm{mS\,cm^{-1}})$, microprocessor controlled and alarm system

Table 10.3. *Industrial instrumentation for single-parameter measurement of dissolved oxygen*

Supplier	Model No.	Description
PHOX	9090	Water-resistant DO meter with recorder output
	9090 HL	Water resistant DO meter with recorder and Hi–Lo alarm and relay outputs
Horiba	WAXA-100	DO monitor with microprocessor and recorder

in a wide range of different types of samples for water-quality monitoring and control.

The integral microprocessor provides automatic temperature compensation, discrimination between zero and span calibration values, salinity compensation and range selection and a self-diagnostic function.

This instrument is intended for on-site monitoring of dissolved oxygen at sewage-processing plants, fish farms, boiler plants and pollution-monitoring stations.

PHOX also supply panel-mounted model PHOX 60 and weatherproof (model PHOX 65) dissolved oxygen recorders and also supply digital multi-channel dissolved oxygen control panel for use in fish farms.

10.1.4 Colour

Kent Industrial Instruments supply the model 8072 on-line colour monitor. This conforms to EEC directive 80/778/EEC by filtering the samples to 0.45 μm and using a measuring wavelength of 400 nm. It is designed for the measurement of water colour on raw water intakes and post-flocculation and activated carbon addition samples. It was developed to meet the demand from the water industry for the measurement of colour on abstracted and potable water.

10.1.5 Turbidity

Horiba produce a model WATA-100 turbidity monitor for industrial waste and potable water. This is an industrial turbidity monitor designed to measure the concentration of the various particulates suspended in water in milligrams per litre (kaolin) or FTU (formazine).

The integral microprocessor automatically selects the ideal measurement range and rapidly performs the complex calculations necessary for accurate turbidity data presentation. Four switched ranges cover a wide range of measurement requirements and make it easy to monitor even low turbidity samples such as tap water. A standard calibration slide made of smoked glass gives fast, foolproof span calibration. Operation is a simple matter of pushing buttons on the front panel.

10.1.6 Miscellaneous

PHOX supply single parameter, panel indicator/controllers for the measurement of temperature (Model PHOX 30), flow (model PHOX 80) and sludge blanket detection (Model PHOX 20). They also supply weatherproof transmitter/indicator controllers for the measurement of these three parameters (respectively models PHOX 35, series 155 and PHOX 25).

10.1.7 Free chlorine

The model 924 Xertex Industrial chlorine measurement system manufactured by Delta Atlantic Corporation is based on an amperometric-sensitive technique. It measures (a) hypochlorous and (b) hypochlorous and plus hypochlorite (total free chlorine) manually if pH is known and controlled and (c) total free chlorine automatically if equipped with a pH sensor. The sensor is protected from direct contact with the sample by a chlorine-permeable membrane. When chlorine permeates the membrane it reacts electrochemically with the cathode and causes a linear current, proportional to the chlorine concentration, to flow between a voltage-based gold cathode and silver anode. The current is then processed in the microprocessor-based analyser transmitter and displayed on an LCD digital display. Applications include potable water treatment, cooling water, process product water, residuals in water mains, cooling towers and industrial cyanide destruction systems. The instrument has built-in self-diagnostics. It is capable of measuring 0–1 and 0–10 mg l^{-1} chlorine in the temperature range 0–50°C and pH range 0–14. The stability is ±0.02 mg l^{-1} chlorine.

10.1.8 Ozone

Delta Analytical Corporation is the only known supplier of industrial equipment for the determination of ozone in water. Their model PDS 940 ozone measurement system is designed for the measurement of ozone in municipal water, industrial water and wastewater plants where ozone is used for disinfection.

The amperometric sensor of the model 940 system consists of gas-permeable membrane, a gold cathode, a silver anode and a buffered salt electrolyte. As ozone gas passes through the sensor's membrane, the gas is reduced at the gold cathode. This reduction process causes a current, proportional to the degree of ozone concentration in water, to flow between the sensor's cathode and anode. The flow of current characterized in the transmitter/analyser's microprocessor electronics is displayed on a 3½ digit LCD display. A thermistor in the sensor provides automatic temperature compensation over the range from 0 to 50°C.

10.1.9 Anions and cations

Industrial scale on-line instrumentation supplied by Kent Instrumentation Ltd for the determination of anions (chloride, nitrate, phosphate, ammonia, fluoride and silica) and cations (sodium, iron) in various types of water are

Table 10.4. *Single-parameter microprocessor-controlled on-line industrial anion and cation analysers from Kent Instrumentation Ltd*

Parameter	Sample types	Model No.	Ranges Concentration	Flow	Temp. (°C)	Pressure (lb in⁻²)	Accuracy	Description	Standardization
Anions									
Chloride	Boiler water water treatment plants	Cabinet version 8024, Panel version 8034	0–100 μg l⁻¹, 0–200 μg l⁻¹, 0–1000 μg l⁻¹, 0–2000 μg l⁻¹, 0–5000 μg l⁻¹	2.5–750 ml min⁻¹	5–45	2–20	± 5 μg l⁻¹ or ± 5%	Solid-state chloride ion-selective electrode and sulphate reference electrode for potentiometric determination of chloride	Auto
Nitrate	Potable water Surface waters	8026	0–1000 μg l⁻¹, 0–500 μg l⁻¹, 500–1000 μg l⁻¹	0.5 l h⁻¹	5–45	2–20	± 5%	Liquid membrane nitrate ion selective electrode	Auto
Phosphate	Water quality monitor reservoirs	8086, Wall mounting 8063, Panel mounting 8064	0–5000 mg l⁻¹, 0–20 mg l⁻¹, 0–10 mg l⁻¹	0.5 l h⁻¹, 0.5 l h⁻¹, 0.5 l h⁻¹	5–55, 5–50, 5–50	2–20, 2–20, 2–20	± 5%, ± 5%, ± 5%	Liquid membrane nitrate ion selective electrode. Spectrometric method. Up to 40 μg l⁻¹ silica results in positive error of ± 2 μg l⁻¹ phosphate error increasing at higher levels of silica	Auto, Auto
Ammonia	Potable water Surface waters	8082	0.05–1000 mg l⁻¹	0.5 l h⁻¹	5–50	2–20	± 5%	Ammonia probe responding to partial pressure of ammonia in sample. Resultant pH change measured by combination glass electrode	Auto
Fluoride	Potable water	8081	0.1–1000 mg l⁻¹	0.5 l h⁻¹	5–55	2–20	± 5%	Fluoride–lanthanum chloride ion-selective electrode with calomel reference electrode sodium	Auto
Silica	Power generation industry	Cabinet mounting 8061, Panel mounting 8062	0–50 μg l⁻¹, 0–200 μg l⁻¹, 0–500 μg l⁻¹, 0–1000 μg l⁻¹, 0–2000 μg l⁻¹	6 ml min⁻¹	5–40	2–20	± 2 μg l⁻¹ or ± 2%	hexametaphosphate–sodium chloride reagents added to suppress aluminium and iron. Spectrophotometric ascorbic acid–H_2SO_4–ammonia citric acid. Reduction up to 5 mg l⁻¹ phosphate causes positive error of < 2 μg l⁻¹ silica 30 mg l⁻¹ phosphate results in positive error of 4 μg l⁻¹ silica	Manual monthly
Sodium	Steam/water	8035	0.1–10.000 μg l⁻¹	16 ml min⁻¹	± 2°C of ambient	with ± 1 meter head of water at inlet flow	± 5%	Electrochemical responsive cell, i.e. sodium-responsive electrode with a calomel reference electrode	Auto
Iron	Potable water	8076	0–100 μg l⁻¹, 0–200 μg l⁻¹, 0–500 μg l⁻¹, 0–1000 μg l⁻¹	5 ml min⁻¹	5–50	2–20	± 2 μg l⁻¹ or ± 2%	Spectrophotometric, complexation of iron with 2,4,6-tri(2-pyridyl)-5-triazine in presence of acetate buffer	Auto

reviewed in Table 10.4. All of these instruments are microcomputer-based and, with the exception of silica, have autocalibration built-in. Generally the sensitivities are adequate for the requirements of the water industry.

In addition Iskra produce an industrial-scale cyanide analyser (model MA 540). This is a sensitive instrument based on the flow electrode measuring cyanide at concentrations down to $1 \mu g\,l^{-1}$ and, as such, might be of interest in the water industry for cyanide monitoring of effluents or reservoir water. PHOX also supply a digital ammonia monitoring module with auto-clearing.

10.2 Multi-parameter instrumentation

Some multi-parameter instruments are reviewed in Table 10.5. The Kent System 19 has been specifically designed for use in the water industry. In these systems continuous single-point and multi-stream analytical instruments gather data for data logging or transmission by telemetry to a control station. Control features of this component are the model 7975 multi-parameter water quality monitor which can analyse up to six streams simultaneously and the 8080 series microprocessor. In abstraction from rivers, reservoirs or boreholes, the instrumentation and systems allow raw water to be continuously monitored before and during pumping to the water treatment plant.

Additionally, mobile or permanent water-quality monitoring stations can be supplied to check environmental conditions and either log the data locally or transmit it by telemetry to a central station. The data thus gathered can be used for control intervention and the generation of reports on, for example, the effects of long-term abstraction.

Both before and during raw water treatment, the instruments and systems play an important role in helping to ensure the quality of the drinking water supply.

A wide range of standard discrete monitoring and control products are offered, many of which are also available in System 19 rack-mounting form for compact and economical installation.

These are ably complemented by other, more specialized equipment and systems for both analytical and water treatment applications, to meet the exacting requirements of the industry.

Kent instruments and systems and water meters are widely used throughout the water distribution network in pipelines, water towers, reservoirs and so on.

In recent years a variety of microprocessor-based and computer-compatible add-on systems have been developed to provide hard data on leakage control, water distribution and other useful analyses.

Table 10.5. *Multi-parameter industrial instrumentation*

Types of sample	Supplier	Model No.
treatment plant intakes and effluents		
(a) Raw water ex rivers, boreholes and reservoirs	Kent Industrial Instruments	Monitor 7975, electronic modules 9516A (range 2–12), 9519 (range 0–1000 electrical conductivity and 0–10 000 μS cm⁻¹) 7995 and 7996 suspended solids, 9527 dissolved oxygen R1072 temperature
(b) Potable water treatment	Kent Industrial Measurements	System 19 instrumentation package for determining level, temperature, flow, pH, dissolved oxygen, conductivity, turbidity, colour, nitrate, ammonia, chloride, silica, fluoride and phosphate
(c) Effluent and sewage treatment	Kent Industrial Measurements	Nitrate, fluoride, ammonia, colour, iron, pH, redox, turbidity, flow, level
(d) Distribution water, water towers, etc.	Kent Industrial Measurements	pH, redox, dissolved oxygen, ammonium nitrate, turbidity, suspended solids, methane, flow, level
	Kent Industrial Measurements	pH, conductivity, colour, iron, turbidity, nitrate, flow, level
Surface water, water treatment, potable water, cooling water, sewage effluent, sea water	Skalar	SA9000 on-line process analyser, aluminium, alkalinity, ammonia, chloride, chlorine, chromium, colour, copper, full cyanide, formaldehyde, hardness, iron free, iron total, magnesium, manganese, nitrate, nitrite, organic carbon, pH, phenol, silica, sulphate, sulphide, sulphur dioxide
Waste water treatment processes	Skalar	Series 8100 process analyser, chloride, fluoride, nitrite, phosphate, bromide, nitrate, sulphate, silica, lithium, sodium, ammonium, potassium, magnesium, calcium, hydrazine, iron, copper, nickel, zinc, cobalt, cadmium, manganese, chromium, lead

As effluent and sewage treatment becomes increasingly subjected to legislation, so water quality monitoring assumes a corresponding important role.

The System 19 instrumentation and systems can be found in all areas monitoring the outfalls from industrial processes, in rivers and at the treatment plant itself.

From individual monitors to complete on-line systems these analysers are deployed throughout many treatment processes to improve efficiency and ensure effective treatment as well as to protect the environment and ultimately the drinking water supply.

The Skalar SA 9000 on-line process analyser is available in two models. As a free-standing unit complete with its own storage section for reagents or as a unit which may be bench or wall mounted or built into an even more comprehensive control system. The main unit consists of the on-line analyser, in which the analysis of the process stream is fully automated. The analysis process is based on the automated addition of sample to reagents which form a coloured complex under controlled conditions. The complex is measured photometrically. The signal is fed into the microprocessor which offers the flexibility for the specific application. Sampling is carried out directly from the process stream. For analysing the same process stream component from several sources, a multi-stream attachment can be supplied. For samples carrying suspended particles, a special continuous in-line filter system is available. Calibration valves allow the introduction of standards and blanks for automatic calibration. Some applications of the SA-9000 analyser are shown in Table 10.6.

The Dionex series 8100 process analyser employs ion chromatography, high-performance liquid chromatography, or flow-injection techniques for the on-line determination of a variety of constituents in process streams. This is a modular instrument which can be modified should processing needs change.

Due to the modularity, the series 8100 can be used in a batch-sampling mode where an operator manually collects the saᵐ ple, or in an on-line mode where the analyser controls all sample selection and sample pretreatment. A series 8100 is easily utilized in an automated batch mode and can be completely upgraded to on-line operation at any time in the future.

A sample preparation module provides automatic dilutions of samples in the range of 1/10 and 1/5000 with precisions normally better than 2% (Figure 10.1). Up to six multi-level calibration standards may be automatically prepared from a stock standard for calibration of the analyser.

Kent supply the series 1800 continuously operating ion-selective electrode monitor which has been used for monitoring concentrations of ammonia, fluoride, nitrate amongst other determinands. These instruments have a microprocessor-based logging system and are equipped with alarm facilities.

Table 10.6. *Applications of Skalar SA-9000 on-line process analyser*

Process stream component	Surface water	Water treatment	Potable water	Boiler water	Cooling water	Condensate	Sewage effluent	Sea water
Alkalinity	●		●					
Aluminium	●		●					
Ammonia	●		●				●	●
Chloride	●		●	●	●	●	●	●
Chlorine	●		●	●	●		●	
Chromium	●		●		●			
Colour	●		●					
Copper	●		●			●		
Cyanide (free)	●		●				●	
Cyanide (total)	●		●				●	
Fluoride	●		●				●	●
Formaldehyde								

Hardness (total)

Iron (free)

Iron (total)

Magnesium

Manganese

Methanol

Nickel

Nitrate

Nitrite

Organic carbon

pH

Phenol

Phosphate

Silica

Sulphate

Sulphide

Sulphur dioxide

Zinc

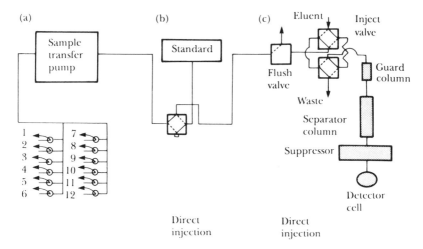

Figure 10.1 *Flowchart: Dionex series 8100 process analyser: (a) sample selection module – up to 18 sample points, continuous or sample on demand, non-metallic, NEMA 4 enclosure; (b) sample preparation module: dilution, reagent addition, standard preparation, preconcentration, matrix elimination; (c) chromatography hydraulics module: ion chromatography, HPLC, flow injection analysis*

10.3 On-line trace metals analyser

Chemtronics Ltd, Australia, have recently introduced two instruments for measurement of microgram per litre levels of toxic metals in water samples. These are the portable digital voltammeter PDV 2000 (discussed earlier) and the on-line voltammetric analyser OVA-2000. Both operate on the electrochemical principle of anodic stripping voltammetry which is very sensitive and specific for trace metals, has low power requirement and is relatively inexpensive.

10.4 Organic pollution monitors

The Horiba OPSA-100 organic pollutant monitor was designed for the continuous measurement of organic pollutants in rivers, lakes and industrial wastewater. No chemical treatment of the sample is necessary. Good correlations are obtained with COD values.

DKK supply the models ODL-12 and SO DL-12 oil-in-water alarm systems. The purpose of these laser instruments is automatic detection of floating oil film and consequent early discovery of oil leakage.

Yorkshire Water Authority, in conjunction with Kontron Ltd, have developed an automated on-line UV pollution monitor for use as a raw water intake protection system at treatment plants.

The method of detection is based on the absorption of UV light (first-order derivatization) by the organic pollutants, using a Kontron Unikon model 860 UV/visible spectrophotometer equipped with a quartz cell under the control of a microcomputer. The system is fully automated, providing 24 hour continuous monitoring for a wide range of organic pollutants, with the facility for signalling an external alarm under exceptional conditions. Phenols, chlorophenols, insecticides, herbicides, hydrocarbons and pyridine have all been determined at sub mg l^{-1} levels with this equipment.

Cognet *et al.* (1987) have discussed the use of ammonia, phenols, heavy metals, hydrocarbons and turbidity monitors on a French water treatment plant located at Le Mont Valerien.

10.5 Applications of telemeters to on-site and on-line analysers

A telecontrol system designed by ATS Telemetry Ltd forms the first phase of an overall telemetry scheme for Yorkshire Water's North and East Division.

At the master station in Harrogate operator interface to the system is by means of two colour monitors with keyboards: information from out-station sites – on reservoir levels, chlorine content, pressures and flows – is displayed, as also are alarm conditions which are automatically printed out to provide a permanent record.

Wilkes (1984) has discussed the use of telemetry in the measurement and control of dissolved oxygen in activated sludge processes.

11 *On-site safety instrumentation*

In this chapter portable instrumentation primarily intended for the personal safety of the operator will be considered, as opposed to instruments intended to monitor the processes. Such instruments monitor the atmosphere for toxic, flammable or asphyxiating gases or a combination of such gases. However, several instruments of this type have a wider application and may be used for atmospheric monitoring and process monitoring provided that the limitations of the instrument are known and understood.

The majority of these instruments use one, or a combination, of the following techniques to monitor the atmosphere.

11.1 Types of gas sensors

11.1.1 Electrochemical cells

Here the gas to be monitored is reacted within the cell with an electrolyte to produce a current in the same way as a galvanic cell. This current is used to produce a reading or to trigger an alarm at preset levels. Until recently cells were only available for a limited range of gases but recent developments have widened this range considerably. The most common use for this type of cell is to measure a gas which is not normally a danger because of its presence but because of its absence, namely oxygen. Most common oxygen monitors within the water industry, are set to alarm at 19% oxygen and are normally used in confined space work. More recent monitors such as the Neotronics Exotox range, the MSA Gas guard and Draeger Triowarn also alarm at 23% oxygen in accord with the Water Research Centre specification for three-in-one instruments. This recognizes the increased use of oxygen within the water industry, on activated sludge plants for example, and the danger of high levels of oxygen being generated due to leakage from an oxygen line or excess levels of oxygen being generated in normal oxy-acetylene cutting operations within a confined space. Such excess levels would not appear to have a harmful effect on personnel but the increased fire risk and the more severe results of any fire do present a real risk.

Other gases monitored by this technique include hydrogen sulphide, carbon monoxide, chlorine and sulphur dioxide. This type of cell has the advantages of being compact, a particular advantage in a personal instrument, easily produced and relatively cheap in manufacture. The disadvantages include cross-sensitivity between gases and perhaps the most serious, the limited life of the cell resulting from the fact that during the reaction the electrolyte is consumed. The rate of this consumption is easy to predict but will vary according to the original specification of the cell and the particular gas being monitored. All manufacturers state a 'life' for their cells and it is vital that these are adhered to as the decay of this type of cell is not linear but tends to follow a gentle slope during the predicted life followed by an unpredictable collapse which has been known to occur in days or hours.

The type of instrument covered in this chapter does not normally fail safe on the toxic gases monitored and does not alarm when the cell fails, although the latest Exotox 60-75 series will alarm for physical failure of the cell. The reason for this failure to alarm is that as far as the instrument is concerned there is no difference between the zero output of a failed cell and the zero output normally associated with the absence of the toxic gas. It is therefore vital that test gas is applied and calibration checks carried out in accordance with the manufacturer's instructions. When this type of cell is used to monitor oxygen, failure of the cell will be to a fail-safe condition as low or zero output will instigate an alarm.

11.1.2 Catalytic oxidation cells

Within a certain range of concentration flammable gas–air mixtures may explode if a source of ignition is present. The concentration range is called the 'explosive range' and the gas–air mixture is called the 'explosive gas mixture'.

The lower concentration limit of the explosive range is called the 'lower explosive limit' (LEL) and the upper limit of the explosive range is called the 'upper explosive limit' (UEL).

The explosive range is normally expressed as the percentage by volume of flammable gas vapour present.

No matter how small is the value of the explosive range, the effect of the explosion is the same once the gas mixture is ignited. To prevent the explosion hazards from flammable gas or vapour, daily checks on vessels or pipe lines must be carefully taken and sufficient ventilation or periodical maintenance of the atmosphere is necessary.

The catalytic oxidation cell is the most common method of flammable gas detection and is widely used for detection up to the LEL of the gas. The principle of this cell is that a catalytic bead is heated to a preset level; at this temperature any molecules of flammable gas in the surrounding atmosphere will be oxidized on the surface of the bead by atmospheric oxygen. This oxidation generates heat which will be given up to the bead. This in turn will cause the temperature of the bead to rise, altering its electrical resistance, as the bead forms one arm of a Wheatstone Bridge electrical circuit, the resulting imbalance within the circuit can be indicated on a meter or used to trip a preset alarm. The heat generated by a given concentration of a flammable gas is clearly related to the calorific value of the gas and so the output of the instrument must be calibrated to a specific gas, normally methane.

The advantages of this type of cell include its robustness, ability to monitor all hydrocarbon-based flammable gases, its long service life and relative cheapness.

The disadvantages of this type of cell include the fact that in common with all catalytic techniques, the catalyst is subject to poisoning from a range of gases including halons, sulphides (it is worth noting that these instruments are often used in areas where hydrogen sulphide may be present), lead and silicones. Poison-resistant cells are available but these often have the penalty of higher power consumption and reduced battery and cell life.

Again recent developments are overcoming these problems to some extent and Sieger with their new SG 7B Siegistor appear to have produced a practical poison-resistant (but not poison-proof) cell by increasing the surface area of the bead. Another problem is that the cell requires the presence of oxygen for the oxidation to take place: as the level of flammable gas increases the level of oxygen decreases and ultimately the instrument ceases to be a flammable-gas-in-air monitor and becomes an air-in-methane monitor, the resistance of the bead being identical at say 1% methane in air (20% LEL) and 90% methane in air. For this reason this type of cell should not be used when levels of flammable gas above the LEL are foreseeable, i.e. during digester purging or commissioning. This problem is accepted by several manufacturers and they have arranged that when the LEL is exceeded the instrument's display panel ceases to continue counting and indicates that the 100% LEL has been exceeded by giving a specific display.

On the MSA Gas guard the display reads '9.9.9' and on the Neotronics Exotox instrument, the display reads 'EEE'. With the Crowcon three-in-one family, which includes 'badge engineered' products including the Shoreco Kanary and the Sabre Sentinel 3, the display continues to count to 99, 100, 101, 102% LEL etc. However, as the response of the detector does not increase linearly with the increase in concentration of flammable gases, the accuracy of these readings decreases. Basically the instrument is being used beyond its intended range and readings above 99% LEL should be taken as simply indicating that the LEL has been exceeded.

The instruments in the water industry, whilst they will respond to all hydrocarbon-based flammable gases, are calibrated for methane with the audible alarm set to 20% LEL (5% by volume) of methane. As stated earlier the output of the cell is related to the calorific value of the gas and so the readings will be in error for gases other than the calibration gas. For example, an instrument calibrated for methane will indicate approximately 50% LEL when actually exposed to 100% LEL of petroleum spirit (not petroleum, as the additives present will also have an effect).

The instrument will alarm below the LEL of all commonly encountered hydrocarbons but not in several cases until well above a true 20% LEL of the gas. Whilst discussing petroleum, which can be found in the sewerage industry following road accidents or spillages, two points should be borne in mind; the first is that petroleum has an OEL (Occupational Exposure Limit) expressed in parts per million and this will have been exceeded long before an alarm is indicated on a flammable monitor and secondly, the majority of petroleum in use at the present time contains tetraethyl lead which has a poisoning effect on the cell. These two factors for all practical purposes, rule out the use of three-in-one type instruments for monitoring atmospheres containing petroleum vapours.

11.1.3 Thermal conductivity cells

In this type of cell a heated glass bead is mounted in a similar way to the catalytic bead in the catalytic oxidation instruments but in this case the rate at which heat is lost from the bead is registered. The rate of heat loss depends on several factors, including radiation, convection and the thermal conductivity of the gas surrounding the glass bead. The conductivity component of the rise in temperature of the bead, for a given energy input, is inversely proportional to the thermal conductivity of the gases present. This again can, with the use of a Wheatstone Bridge circuit, be used to indicate the gas concentration. The thermal conductivity of methane is higher than that of air, so it follows that as the level of methane increases the temperature of the glass bead will fall. The instrument can therefore be calibrated to record all levels of methane from 0 to 100% by volume. This calibration will not be as fine as instruments working on the catalytic oxidation principle over the 0–100% LEL (0–5% by volume) range and so several instruments are dual scale using both types of cell, catalytic oxidation for the LEL scale and thermal conductivity for the percentage volume scale.

The advantages of this type of cell are its ability to record the full 0–100% volume scale, its relative resistance to poisoning and the ease of calibration to any known gas mixture. The latter advantage is of particular use in monitoring the performance of sewage digesters as the instrument, if calibrated for methane in carbon dioxide, will give direct on-site readings of the quality of gas output from a digester. This can be of great benefit when combined heat and power (CHP) units are in use. Also the instrument has a long service life and is easily calibrated using standard calibration gases.

The disadvantages are more complex and often misunderstood, frequently leading to mis-application of the instruments. For practical purposes the thermal conductivity of air, which contains 79% nitrogen, can be taken to be the same as pure nitrogen. So an instrument calibrated for methane in nitrogen can be used to measure methane in air. However, the thermal conductivity of carbon dioxide is considerably lower than that of nitrogen and an instrument calibrated for methane in nitrogen will read low if it is used to measure methane in carbon dioxide. The reverse is also true and an instrument calibrated for methane in carbon dioxide will over-read a mixture of methane in nitrogen. It is therefore clear that the actual gas mixture to be measured must be known if valid results are to be obtained. For example a normal healthy digester gas contains a mixture of methane in carbon dioxide and only thermal conductivity instruments calibrated for this mixture will give valid results. However, the usual gas used to render digesters inert is nitrogen and during a purge the atmosphere within the digester will progressively change with the mixture of methane and carbon dioxide being replaced by nitrogen. Under these circumstances the instrument calibration will correspondingly become more inaccurate and, as the levels of methane required to indicate a successful purge are low, say 2% by volume, and, this is the area where the instrument is at its most inaccurate, this type of instrument is not suitable when purging with nitrogen. It would, however, be well suited when the purge gas is carbon dioxide. In order to overcome this problem two avenues are open. Firstly, when purging with nitrogen, instruments calibrated for methane in nitrogen may be used on the basis that the amount of carbon dioxide present in the final stages of the purge is so low that the inaccuracy in the readings it will cause are not practically relevant. However, it is theoretically possible that a combination of methane and carbon dioxide will cancel each other out and produce a depressed or zero reading. The danger is that the instrument could indicate that the purge had been completed when in fact a significant quantity of methane remained within the digester; air would then be introduced with possible disastrous results. The alternative approach is to incorporate a device within the sampling line to remove the carbon dioxide present and therefore remove the risk of any false readings. Such a device is easily assembled, and consists of a length of clear plastic tubing of approximately 20 mm bore and 300 mm length filled with granular carbon dioxide absorber and glass wool plugs and adaptors at each end. The adaptors are to enable the device to be fitted to the smaller-bore sample line. Carbon dioxide absorbers are now available which change colour as they are used up and this change in colour can be readily seen through the clear plastic. The removal of the carbon dioxide will slightly reduce the volume of the sample and therefore enhance the methane reading but the over-reading is slight and the results are acceptable.

The principal point is that for thermal conductivity monitors to be used satisfactorily, their limitations must be understood and the nature of atmosphere known, i.e. methane in carbon dioxide or methane in nitrogen, in order that the most suitable monitor can be chosen. A problem arises when air has been accidentally introduced into a digester, either by mechanical failure or operator error. In this case the degree of air ingress must be determined before a thermal conductivity methane monitor is used as this will indicate the type most appropriate to the atmosphere present within the digester. A simple way is to monitor for oxygen, ensuring that the monitor used is suitable for an atmosphere containing carbon dioxide. As air contains a ratio of 4 to 1 nitrogen to oxygen this reading will indicate whether the atmosphere is predominantly nitrogen or carbon dioxide and the most appropriately calibrated monitor can be chosen. Plotting methane against oxygen will indicate whether an explosive atmosphere is present.

11.1.4 Infrared detectors

When the gas to be analysed does not contain oxygen (e.g. digested gases containing only nitrogen and methane), then the catalytic oxidation type of cell is not appropriate to the determination of methane, because oxygen is absent in the sample. Similar comments apply in the case of oxygen-free mixtures of methane and carbon dioxide. For such mixtures methods based on thermal conductivity or infrared are applicable to the determination of methane. Infrared detectors are based on the absorption, by methane, of infrared radiation. The amount of absorption is measured and converted into a concentration of methane.

11.1.5 Mechanical/flame detectors

The remaining type of on-site instrumentation, that of the miner's flame lamp, will be touched on only briefly. A flame of a preset height using a specific fuel will decrease in height if the oxygen level is reduced and will be extinguished if the oxygen content falls below a certain level. The same flame will burn higher and give off more heat if a flammable gas is present. This principle has been used in both the mining and sewerage industries for many years. The latest of the detectors based on this simple principle have the flame set so that a reduction of the oxygen level to below 19% extinguishes the flame and the presence of 20% LEL. Methane causes the flame to produce sufficient heat to trigger a microswitch by the movement of a heat-sensitive bimetal coil. This microswitch operates a buzzer which warns the operative. Detection of hydrogen sulphide is achieved by moistening a piece of filter paper with lead acetate solution, which stains in the presence of hydrogen sulphide. The drawbacks are as follows: the operator sets the detector each time it is lit, leading to inaccuracies in the setting of the flame height and therefore the effectiveness of the detector; the unit takes an hour to reach operating temperature; no audible alarm is given for either oxygen deficiency or hydrogen sulphide; and recalibration is required every six weeks. These drawbacks make this type of detector impractical for present day use.

11.2 Instrument checking and calibration

All the detectors and monitors referred to in Section 11.1 require checking and calibration, which must be carried out in accordance with the manufacturers' instructions at the appropriate intervals. A clear distinction must be made between checking and calibration.

Checking is normally carried out monthly and involves sampling into the instrument test gases and checking its response. If the instrument reacts and alarms, it has passed the check and can be returned to service provided no physical damage is apparent. This type of check involves no resetting or calibration, and several instrument manufacturers (e.g. Neotronics and M Tec) produce test gas kits to enable these tests to be carried out on-site by trained operatives. Records of these tests should be available with the instrument.

Calibration is more involved as the time taken for cells to react and the output of the cells have to be taken into account. Most manufacturers work on a three- or six-monthly recalibration period when either trained in-house technicians or manufacturer's service staff carry out the calibration of the instrument.

11.3 Types of instruments available and factors affecting their selection

11.3.1 Factors affecting instrument selection

Great care should be taken in the selection of gas monitoring/detection instruments. The primary concern is that the equipment is capable of detecting the foreseeable hazardous atmospheres; most 3 in 1 detectors will monitor only one toxic gas (normally hydrogen sulphide) and a second supplementary detector may be required for other foreseeable toxic gases. The detector must also be fit for its purpose. The Water Research Centre (UK) carries out evaluation of portable gas monitors to a specification laid down by a technical group drawn from within the water industry and manufacturers. The evaluation document is available from WRC Engineering, Frankland Road, Blagrove, PO Box 85, Swindon, Wiltshire, SN5 8YR. It is not proposed to publish a summary here as the specification and detectors covered are constantly under review. The environment in which the equipment is to be used must also be considered. Equipment will in all probability require to be certified as instrinsically safe in accordance with British Standard No. 5345, Selection, Installation and Maintenance of Electrical Apparatus for use in Potentially Explosive Atmospheres, Part 110. It should be noted that in various publications sewers are classified as zone 0 (where an explosive gas–air mixture is assumed to be continuously present or present for long periods), zone 1 (where an explosive gas–air mixture is likely to occur in normal operations) or zone 2 (where an explosive gas–air mixture is not likely to occur in normal operation and if it does occur it will exist only for a short time). The sewerage system is so large and varied that there is validity in all these classifications; however, the locations in which the zone 0 applies are extremely rare and do not occur in the normal operation of the sewerage system. Some short lengths of sewer associated with the drainage of tip leachate or spillages from chemical or petrochemical plants can prove the exception to this rule. It is the author's experience over 20 years, that flammable atmospheres occur very rarely and the absence of explosions within the sewerage systems appears to bear out this anecdotal experience leading to the view that for the 'normal' sewerage system zone 2 would be most appropriate. However, the possible risks of spillages from industry, road transport, rail transport, ground conditions and blockages leading to septic conditions over which there is little control leads to the conclusion that equipment appropriate for use in a zone 1 methane atmosphere is the most suitable for use in the water industry. An added advantage of equipment to this classification is that it is also appropriate for use in association with digestion plants.

Other considerations include the high humidity and moisture, the presence of grease and fats on the walls of manholes which easily blocks the cells, the all-up weight of the equipment (this may not be a problem if the detector is not to be carried) and the rough handling which inevitably occurs. All of these factors combine to make the choice of instrument a compromise. As a general rule the lighter a detector the less robust it will prove and the lighter an instrument the greater the user acceptability.

11.3.2 Instrument suppliers

Crowcon gas testers

This company supplies two instruments suitable for analysing gas mixtures in hazardous areas, e.g. sewer networks. The Crowcon gas tester type 75 TC and the Crowcon gas tester triple type 84 TR.

Crowcon gas tester type 75 TC
This instrument uses a thermal conductivity sensor to measure the composition of known gas mixtures (see Section 11.1.3). Its operation relies on the fact that the thermal conductivities of gases differ and if the difference between two particular gases is sufficient, reliable measurement can take place. It uses a manual aspirator to take samples of the gas mixture.

Instruments for the determination of the concentrations of gases shown in Table 11.1 are available.

Table 11.1. *Gas and concentration ranges determined by the Crowcon 75TC instrument*

Gas mixture	Concentration range (%)					
	0–1	*0–5*	*0–10*	*0–25*	*0–50*	*0–100*
Carbon dioxide in air		●	●	●	●	●
Helium in air/nitrogen	●	●	●	●	●	●
Hydrogen in air/nitrogen		●	●	●	●	●
Methane in air/nitrogen			●	●	●	●

The 75 TC can be calibrated for other gas mixtures to suit various applications. For example an instrument calibrated 90–100% helium in air can be used as a helium purity comparator.

Crowcon Gastester 84 TR

This instrument (also sold as the Shorec Kanary and the Sable Seninel Three) is applicable where the simultaneous monitoring of three gas hazards is essential.

The 84 TR has a pellistor sensor for flammable gas detection (calibrated for methane, pentane or other saturated hydrocarbons) an electrochemical toxic gas sensor (for hydrogen sulphide or optional carbon monoxide) and an oxygen sensor cell (providing warning of low or high levels of oxygen).

The triple 84 TR has been approved by BASEEFA as intrinsically safe, the pellistor detector meeting special requirements. It is approved as type ia and can therefore be used in zones 0, 1 or 2. It provides maximum safety in all potentially hazardous areas, even inside storage tanks containing flammable liquids.

Should any one of the three gas hazards occur, the appropriate alarm lamp will light and the sounder will operate, irrespective of the position of the control knob. The alarm can only be reset when the gas concentration is below the trip level.

M-TEC combustible gas detector alarm

This combustible gas detector contains a pair of filaments matched as to their electrical characteristics. The active filament has catalytic response to the presence of combustible vapours with a corresponding increase in temperature and resistance. The other filament serves only as a thermal reference. This detector is therefore based on the catalytic oxidation principle discussed in Section 11.1.2. The electrical difference between them is compared in a Wheatstone bridge and then suitably amplified for indication and alarm. Both the active and reference filaments are contained in a single sintered stainless steel flame arrestor which is safe in any gas–air mixture.

The Draeger Triowarm

This instrument continuously monitors the atmosphere for oxygen, flammable and toxic gases (hydrogen sulphide and carbon monoxide). In addition, it shows levels of all three gases simultaneously on the display panel and triggers visual and audible alarms at pre-set levels. The instrument features electrochemical detectors, automatic fault finding and fail-safe operation and is designed to meet the requirements of British Standard BS 6020.

Neotronics Exotox range of gas monitors

Neotronics offer a range of intrinsically safe gas monitoring instrumentation.

This company also supplies a flammable gas detector, the Digiflam, suitable for detecting flammable gas levels, e.g. in landfalls, sites and sewers.

Exolox Model 50 four-in-one gas monitor

This is a four-gas monitor for oxygen deficiency/enrichment, flammable gas and toxic gas measurements, choice of any two toxic gases from carbon monoxide, hydrogen sulphide or sulphur dioxide.

Exotox Model 20/25/40 three-in-one gas monitor

This is a three-gas monitor, i.e. as above but only one toxic gas included. The instrument has a go/no go alarm and a continuous flammable gas sensor.

Exotox Model 55 four-in-one gas monitor with GL-15 logger

This is a four-gas monitor with an LCD real-time display of results, full data logging capability for recording and storing information for long-term analysis.

Exotox 60, 70 and 75 five-gas monitors

All three of these instruments provide rapid measurement of gas levels, a clear, digital indication of the safety of any work area, and can be relied upon to give prompt warning of developing gas hazards.

The monitors can be fitted with up to five gas sensors, catalytic and electrochemical gas sensors for oxygen, flammable and toxic gases in addition to temperature and relative humidity, providing comprehensive environmental monitoring. With the increased number of gas sensors fitted to the Exotox this new range represents a major advance in gas monitoring capability.

IGD Sentinel three-in-one gas detector

This is a three-gas detector, based on the catalytic oxidation detector cell, capable of determining flammable gases (methane) in the 0–100% LEL range, hydrogen sulphide (0–50 p.p.m.) and oxygen (10–30%). Alarm settings are at 20% LEL (methane), 10 p.p.m. (hydrogen sulphide) and 10–30% (oxygen). Visual and audible gas alarms are available. The instrument will operate for up to eight hours continuous monitoring and can also be used for random testing of the atmosphere.

MSA Gasguard 742 three-in-one gas monitor

This is another portable three-in-one gas monitor designed to meet British Standard BS 6020, featuring electrochemical and catalytic oxidation detection cells for the determination of methane, carbon monoxide and oxygen.

If any of the gases to be measured are present in concentrations exceeding the pre-set alarm levels, the audible alarm will sound and the appropriate visual indicator will flash. These alarms are latched and can only be reset if the gas levels decrease or the instrument is removed from the gas hazard area.

Neotronics Neotox pocket sized single gas monitors

This is a range of pocket-sized monitors for protection against a single gas–oxygen deficiency and enrichment, carbon monoxide, hydrogen sulphide, sulphur dioxide or chlorine.

Sieger Gas Leader four-in-one monitor

This instrument is capable of continuous monitoring of one to four different gases from the list methane, hydrogen sulphide, oxygen deficiency, carbon monoxide, sulphur dioxide and chlorine. Cartridge sensors for hydrogen, nitric oxide and nitrogen dioxide are also available. The methane monitor is based on an infrared detector and consequently this detector is applicable to the determination of methane in oxygen-free methane–nitrogen and methane–carbon dioxide mixtures.

The Gas Leader has fully adjustable alarm levels, and gas levels above these limits activate the integral audible and visual alarms menu-driven software governs calibrations and instrument adjustments. A two-way communication point is built in to allow the transfer of information to a data logger enabling hard copy historical data of gas level trends within working areas to be made, stored and updated.

In Table 11.2 is summarized information on the gases that can be determined by each of the instruments discussed above.

Table 11.2. *Single gas-testing instruments – multiple gas-testing instruments*

	Single gas-testing instruments									Multiple gas-testing instruments							
	CO_2	O_2	He	CO	Cl_2	H_2	H_2S	CH_4[1]	SO_2	$CH_4/$ $H_2S/$ O_2	$CH_4/$ $CO/$ O_2	$CH_4/$ $CO/$ $H_2S/$ O_2	$CH_4/$ $O_2/$ LEL	$CH_4/$ 2 toxics/ O_2	$CH_4/$ 1 toxic/ O_2	$CH_4/$ $CO/$ $SO_2/$ $H_2S/$ O_2	$CH_4/H_2/$ $H_2S/$ $NO/O_2/$ $NO_2/$ SO_2/Cl_2
Crowcon 75 TC	✓ in air		✓ in air/ nitrogen			✓ in air/ nitrogen		✓ in air/ nitrogen									
Crowcon 84 TR										✓	✓						
H-Tec								✓									
Digiflow Draeger Triowarm												✓					
Neotronics Digiflam 2000								✓		✓							
Exotox 50														✓			
Exotox 20/25/40															✓		
Exotox 55																	
Exotox 60														✓		✓	
Exotox 70																✓	
Exotox 75																✓	
Neotox pocket-sized single-gas monitors		✓		✓	✓		✓		✓								
Sieger gas header													✓				
GMI Landsurvey 1										✓							✓
IGD Sentinel 3											✓						
MSA Gasguard 742																	

[1] CH_4 – flammable gases.
[2] Choice of two from CO, SO_2, H_2S.
[3] Choice of one from CO, SO_2, H_2S.

Appendix
Instrument suppliers

Sections 2.1 and 2.2

Flame and graphite furnace atomic absorption spectrometry

IL and Video Series:
Thermoelectron Ltd (formerly Allied
Analytical Systems)
830 Birchwood Boulevard
Birchwood
Warrington
Cheshire WA3 7QT
UK

Thermoelectron Ltd
590 Lincoln Street
Waltham
MA 02254
USA

Perkin-Elmer 2280, 2380, 1100 and 2100
Perkin-Elmer Ltd
Post Office Lane
Beaconsfield
Bucks HP9 1QA
UK

Perkin-Elmer Corporation
Analytical Instruments Division
761 Main Avenue
Norwalk
Connecticut 06856
USA

Varian Associates Spectr AA 30/40 and Spectr
AA 10/20
Varian Associates Ltd
29 Manor Road
Walton on Thames
Surrey KT12 2QF
UK

Varian Techtron Pty Ltd
679 Springvale Road
Mulgrove
Victoria
Australia 3170

Varian Instruments Division
611 Hansen Way
Palo Alto
California 94303
USA

Varian AG
Steinlausertrasse CH-6300
Zug
Switzerland

GBC 903 and 902
GBC Scientific Equipment Pty Ltd
22 Brooklyn Avenue
Dandenong
Victoria
Australia 3175

UK Agent
Techmation Ltd
58 Edgeware Road
Edgeware
Middlesex HA9 8JP
UK

Shimadzu AA 670 and AA 670G
Shimadzu Corporation
International Marketing Division
Shinjuki Mitsui Building, 1-1
Nishe-Shinjuku 2 Chrome
Shinsuku-ku
Tokyo 163
Japan

UK Agent
V A Howe Co Ltd
12–14 St Ann's Crescent
London SW18 2LS
UK

Autosamplers

20.020 20 Position autosampler
20.080 80 Position autosampler and fraction
collector atomic absorption model comprising:

PSA 20.080 SS Stainless steel probe
PSA 20.080 PP Polypropylene probe
PSA 20.080 CC Complete automatic control
from computers
PSA 20.080 OA On-line dilution probe
PSA 20.080 FC Triple-probe fraction collector
PSA 20.080 VP Vial piercing option
PSA 20.080 PH PH Electrode assembly
PSA 20.080 TM Turrax mixer assembly
   and interface requirements:
TTL logic
RS 232 Random access
RS 232 Other requirements
   PS Analytical Ltd
   Arthur House
   Cray Avenue
   Orpington
   Kent BR5 3TR
   UK

Gilson 300 position programmable
autosampler
   Gilson International
   Box 27
   300 W Beltine
   Middleton
   Wisconsin 53562
   USA

   Gilson Medical Electronics (France) SA
   BP 45 F-95400
   Villiers-le-Bel
   France

Section 2.3
Zeeman atomic absorption
spectrometry

Perkin-Elmer Zeeman 3030 and Zeeman 5000
   Perkin-Elmer Ltd
   Post Office Lane
   Beaconsfield
   Bucks HP9 1QA
   UK

   Perkin-Elmer Corporation
   Analytical Instruments Division
   761 Main Drive
   Norwalk
   Connecticut 06856
   USA

Varian Associates Spectr AA30/40 and Spectr
AA 300/400
   Varian Associates Ltd
   28 Manor Road

Walton on Thames
Surrey KT12 2QF
UK

Varian Techtron Pty Ltd
679 Springvale Road
Mulgrove
Victoria
Australia 3170

Varian Instrument Division
611 Hansen Way
Palo Alto
CA 94303
USA

Varian AG
Steinhauserstrasse CH-6300
Zug
Switzerland

Section 2.5
Hydride generator
assemblies

PS Analytical Ltd Model PSA 10.002
   Dr P B Stockwell
   PS Analytical Ltd
   Arthur House
   Far North Building
   Cray Avenue
   Orpington
   Kent BR5 2TR
   UK

Varian AG VGA-76
   Varian Associates Ltd
   28 Manor Road
   Walton on Thames
   Surrey KT12 2QF
   UK

   Varian Techtron Pty Ltd
   679 Springvale Road
   Mulgrove
   Victoria
   Australia 3170

   Varian Instrument Division
   611 Hanson Way
   Palo Alto
   CA 94303
   USA

   Varian AG
   Steinhauserstrasse CH-6300
   Zug
   Switzerland

Section 2.6
Inductively coupled plasma optical emission spectrometers

Spectroflame
  Spectro Analytical UK Ltd
  Fountain House
  Great Cornbew
  Halesowen
  West Midlands B63 3BL
  UK

  Spectro Inc
  160 Authority Drive
  Fitchbury 01420
  Massachusetts
  USA

PU 7450 and PV8050
  Phillips Analytical Agents, Pye Unicam Ltd
  York Street
  Cambridge CB1 2PX
  UK

  Phillips Electronic Instruments
  85 McKec Drive
  Mahwah
  New Jersey 07430
  USA

Miscellaneous instruments

Perkin-Elmer Ltd
Post Office Lane
Beaconsfield
Bucks HP9 1QA
UK

Perkin-Elmer Corporation
Analytical Instruments Division
761 Main Avenue
Norwalk
Connecticut 06856
USA

Plasma test system 75
  Labtest Equipment Co
  11828 Grange Avenue
  Los Angeles CA 90025
  USA

  Labtest Equipment (Europa) GmbH
  4030 Ratinger

Tal Strasse 35
Ratingen
Germany

Spectrovac
  Baird Atomic Ltd
  4 Warner Drive
  Springwood Industrial Estate
  Braintree
  Essex CM7 7YL
  UK

Plasma 200 and Plasma 300
  Thermoelectron Ltd (formerly Allied
  Analytical Systems)
  830 Birchwood Boulevard
  Birchwood
  Warrington
  Cheshire WA3 7QT
  UK

  Thermoelectron Ltd
  590 Lincoln Street
  Waltham
  NA 02254
  USA

Labtam 8440 and Plasmascan 8410
  UK Distributors
  Techmation Ltd
  58 Edgeware Way
  Edgeware
  Middlesex HA8 8JP
  UK

  AB Labtam Ltd
  43 Malcolm Road
  Braeside
  Victoria
  Australia 3195

Section 2.7
Inductively coupled plasma mass spectrometers

Plasmaquad
  VG Isotopes Ltd
  Ion path, Road three
  Winsford
  Cheshire
  UK

Elan 500
  Perkin-Elmer Ltd
  Post Office Lane
  Beaconsfield
  Buckinghamshire HP9 1QA
  UK

Auto samplers
Intelligent autosampler 84100
R 5232 C serial
Communications interface

A B Labtam Ltd
43 Malcomb Road
Braeside
Victoria
Australia 3195

Section 2.8
Inorganic mass spectrometry

Model 251 Isotope ratio mass spectrometer
Model 281 UF6 mass spectrometer
Models 271 and 271/45 Hydrocarbon group type mass spectrometer
Thermionic quadrupole mass spectrometer
Model 261 Magnetic sector mass spectrometer
Delta and Delta E stable isotope ratio mass spectrometers
Delta isotope ratio mass spectrometer

Finnigan MAT
355 River Oaks Parkway
San Jose
California
CA 95134-1991
USA

Finnigan MAT Ltd
Paradise
Hemel Hempstead
Herts HP2 4TQ
UK

Isochrom II
Gas chromatograph–infrared-mass spectrometer for isotope ratio mass spectrometry

VG Isogas Ltd
Cheshire
UK

Section 2.9
Visible ultraviolet and near infrared spectrometers

UV, visible and near infrared
PU 8620 basic instrument
Optional PU 8700 scanner for colour graphics
PU 8800 research applications

Pye Unican Ltd

York Street
Cambridge CB1 2PX
UK

Phillips Electronic Instruments
85 McKee Drive
Mahway NJ 07430
USA

Phillips Nederland BU
HSD Analysetechnicken VB3
Postbus 90050
5600 PB Eindhoven
Netherlands

Cecil Instruments Ltd
CE 2343 Visible range spectrophotometer
CE 243D Visible range spectrophotometer
CE 2393 Digital visible grating spectrophotometer
CE 2292 Digital ultraviolet spectrophotometer
CE 2303 Grating spectrophotometer
CE 2202 Ultraviolet spectrophotometer
CE 2373 Linear read-out grating spectrophotometer
CE 2272 Linear read-out ultraviolet spectrophotometer
CE 594 Ultraviolet and visible double-beam spectrophotometer
CE 6000 Ultraviolet visible double-beam spectrophotometer with
CE 6606 Superscan graphic plotter

Cecil Instruments Ltd
Milton Industrial Estate
Cambridge CB4 4AZ
UK

Kontron Instruments
Unikon 860 Ultraviolet visible double-beam spectrophotometer
Unikon 930 Ultraviolet – visible graphics

Kontron Instruments AG
Bernerstrasse, SUD 169
8010 Zurich
Switzerland

Perkin-Elmer Ltd
Lambda 2 Ultraviolet–visible double-beam spectrophotometer
Lambda 3 Ultraviolet–visible double-beam spectrophotometer
Lambda 5 and Lambda 7 Ultraviolet visible spectrophotometers
Lambda 9 Ultraviolet visible – near infrared spectrophotometer
Lambda Array 3430 Spectrophotometer

Perkin-Elmer Ltd
Post Office Lane
Beaconsfield
Buckinghamshire HP9 1QA
UK

Perkin-Elmer Corporation
Analytical Instruments Division
761 Main Avenue
Norwalk, CT 06859
USA

Bodenseewerk Perkin-Elmer P 60 GmbH
Postfach 1120
D-770
Uberlingen
Germany

Section 2.10
Energy dispersive and total reflection X-ray fluorescence spectrometers

Energy-dispersive types
Link Analytical XR 200/300
  Link Analytical Ltd
  Halifax Road
  High Wycombe
  Buckinghamshire HP12 3SE
  UK

  Link Analytical Ltd
  240 Twin Dolphin Drive
  Suite B
  Redwood City
  California 94065
  USA

Phillips PW 1404
  Pye Unicam Ltd
  York Street
  Cambridge CB1 2PX
  UK

Phillips Electronic Instruments
  85 McKee Drive
  Mahway NJ 07430
  USA

  Phillips Nederland BV
  Afd. Analysetechniken VB3
  Postbus 90050
  5600 PB Eindhoven
  Netherlands

Energy dispersive and total reflection types:
Siefert Extra 2
  Richard Siefert & Co
  GmbH & Co KG
  Bogenstrasse 41
  Postfach 1280
  D-2070 Ahrenberg
  Germany

Section 2.12
Polarography, voltammetry

646 VA Processor/647 VA stand/675 VA
sample changer and 665 decimat
506 Polarecord
626 Polarecord
  UK Agents
  VA Howe & Co Ltd
  12–14 St Ann's Crescent
  London SW18 2LS
  UK

  Metrohm Ltd
  CH-9100 Herisau
  Switzerland

ECP 100 Differential Pulse Polarograph
ECP 120 ASW Programmer
ECP 140 anodic stripping analyser
  EDT Analytical Ltd
  14 Trading Estate Road
  London NW10 7LH
  UK

ECP CYSY-1 Computer controlled electro-
analytical system
PDV-2000 portable trace metal analyser
CVA-2000 on-line voltammetric analyser
ECP-100 differential pulse polarograph
  EDT Analytical Ltd
  14 Trading Estate Road
  London NW10 7LU
  UK

  Cypesso Systems Inc
  PO Box 3931
  Lawrence KS 66046
  USA

PDV-2000 on-line voltametric analyser
  Chemtronics Ltd
  Bentley
  Western Australia

Chapter 3
Water purification units

Autostill range
  Jencons Scientific Lab
  Cherrycourt Way Industrial Estate
  Stanbridge Road
  Leighton Buzzard
  Bedfordshire LU7 8UA
  UK

Water 1 column unit
  Gelman Sciences
  10 Horrowden Road
  Brackmills
  Northampton NN0 0EB
  UK

L4 stainless steel water cell
  Manestry Machines
  Speke
  Liverpool L24 9LQ
  UK

48C/24C Deioniser
  Houseman (Burnham) Ltd
  UK Industrial Division
  Waterslade House
  53–57 High Street
  Maidenhead
  Berks SL6 1JU
  UK

Still
  Hamilton Laboratory Glassware
  Europa House
  Sandwich Industrial Estate
  Sandwich
  Kent CT13 9LR
  UK

Aquatron Stills
  J Bibby Science Products Ltd
  Stone
  Staffordshire ST15 0SA
  UK

Nanopure ultra pure water system and Cyclon
advanced ultrapure still and RO60 reverse
osmosis unit
  Fistreem Water Purification
  Belton Road West
  Loughborough
  Leicestershire LE11 0TR
  UK

Elgastat Spectrum
Elgastat UHP
Elgastat UHQ
Elgastat Prima
  Elga Ltd
  Lane End
  High Wycombe
  Buckinghamshire HP14 3JH
  UK

Milli Q System
Milli RO system
  Water Chromatography Division
  Millipore (UK) Ltd
  11–15 Peterborough Road
  Harrow

Middlesex HA1 2YH
UK

Millipore Intertech
PO Box 255
Bedford
MA 01730
USA

Section 4.1
Segmented flow analysers

SAN Plus system
  Scalar BV
  Spinvald
  LL 4815 HS Breda
  PO Box 3237
  NL 4800 DE Breda
  Netherlands

Chemlab CAAI system
Chemlab CAAII system
System 4
  Chemlab Instruments Ltd
  Hornminster House
  129 Upminster Road
  Hornchurch
  Essex RM11 3XJ
  UK

Metrohm 670 Titroprocessor
Metrohm 682 Titroprocessor
Metrohm 686 Titroprocessor
  UK Agent
  V A Howe & Co Ltd
  12–14 St Ann's Crescent
  London SW18 2LS
  UK

  Metrohm Ltd
  CH 9101 Herisau
  Switzerland

Mitsubishi Automatic titration model GT 05
Mitsubishi Automatic sample charger model
GT 05SC
  Mitsubishi Chemical Industries Ltd
  Instrument Dept
  Mitsubishi Building
  52 Marunouchi 7 chrome
  Chiyoda Ku
  Tokyo 100
  Japan

Jenway PCLH digital chloride water
  Jenway Ltd
  Gransmore Green
  Felsted
  Dunmow
  Essex CM6 3LB
  UK

Section 4.3
Flow injection analysis

FIA star 5020
FIA star 5032
FIA star 5025 ion selective electrode meter
FIA star 5010
The Aquatec system
  UK Supplier
  EDT Analytical Ltd
  14 Trading Estate Road
  London NW10 7LN
  UK

  Tecator AB
  Box 70
  S 26301 Hanagas
  Sweden

FIA System LGC1
  Advanced Medical Supplies Ltd
  Caker Stream Road
  Mill Lane Industrial Estate
  Alton
  Hampshire GU34 2PL
  UK

Chemlab FIA system
  Chemlab Instruments Ltd
  Hornminster House
  129 Upminster Road
  Hornchurch
  Essex RM11 3XJ
  UK

FIA
  Skalpar BV
  Spinvald
  LL 4815 HS Breda
  PO Box 3237
  NL 4800 DE Breda
  Netherlands

Fialite 600
Fiatrode 400
Fiatrode 410
  Fiatron Laboratory Systems
  5105 South Wortington Street
  Oconomowoc WI 53066
  USA

Fiazyme 500 Series Automated Carbohydrate
Analysis Models

Section 4.4
Ion-selective electrode equipment

EA 940
EA 920
SA 720
SA 720
Orion 960 Autochemistry System with
optional 960SC sample changer
  Orion Research UK
  Freshfield House
  Lewes Road
  Forest Row
  East Sussex RH18 5ES
  UK

  Orion Research Incorporated
  Laboratory Products Group
  The Schraft Centre
  529 Main Street
  Boston MA 02129
  USA

Ion-selective electrodes
  Ingold Electrodes Ltd
  261 Ballandvale Street
  Wilmington MA 01887
  USA

Ion-selective electrodes
  EDT Analytical Ltd
  14 Trading Estate Road
  London NW10 7LU
  UK

Section 4.5
Ion chromatography

Wescam System single-channel ion
chromatograph, dual-channel ion
chromatograph and dual-channel automated
ion chromatograph
  UK agents
  Alltech Associates Applied Science Ltd
  6–7 Kellet Road Industrial Estate
  Carnforth
  Lancashire LA5 9XP
  UK

  Wescan Instruments Inc
  2051 Waukegau Road
  Deerfield
  Illinois 60015
  USA

6200 Ion Analyser, 6210 pump
  Tecator AB
  Box 70
  S26301 Hanagas
  Sweden

  Tecator Ltd
  Cooper Road
  Thornbury
  Bristol BS12 2UW
  UK

  UK Agent
  EDT Analytical Ltd
  14 Trading Estate Road
  London NW10 7LU
  UK

HIC 6A High performance ion chromatograph
  UK Agent
  Dyson Instruments Ltd
  Hetton Lyons Industrial Estate
  Hetton
  Houghton le Spring
  Tyne and Wear DH5 3RH
  UK

  Shimadzu Corporation
  International Marketing Division
  3 Kanda
  Nisikicho 1 chrome
  Chiyoda ku
  Tokyo 101
  Japan

Ion chem system
  UK Agent
  Severn Analytical
  36 Brunswick Road
  Gloucester GL1 1JJ
  UK

  ESA Inc
  45 Wiggins Avenue
  Bedford MA 01730
  USA

Monitor III ion chromatography system
  LCD Milton Royal
  PO Box 10235
  Riviera Beach
  Florida L 33404
  USA

  LDC UK
  Milton Roy House
  52 High Street
  Stone
  Staffordshire ST15 8AR
  UK

also
  HPLC Technology Ltd
  Wellington House
  Waterloo Street West
  Macclesfield
  Cheshire SK11 6PS
  UK

690 Ion chromatograph
  Metrohm Ltd
  CH 9100 Herisau
  Switzerland

  UK Agents
  V A Howe and Co Ltd
  12–14 St Ann's Crescent
  London SW18 2LS
  UK

Dionex Autoion 400 series 2000i 4000i, 4500i
ion chromatographs
  Dionex UK Ltd
  Selmoor Road
  Farnborough
  Hampshire
  UK

  Dionex Corporation
  PO Box 3603
  Sunnyvale CA
  USA

ILC Series ion chromatograph
  Waters Division of Millipore
  Millipore UK Ltd
  Waters Chromatography Division
  11–15 Peterborough Road
  Harrow
  Middlesex HA1 2YH
  UK

Section 5.1
pH meters

Beta 51, Beta 52, Beta 53, Beta 500, Micro 2,
Alpha 200, Alpha 500, Alpha 600, Sigma 1
  EDT Ltd
  Lorne Road
  Dover
  Kent CT16 2DR
  UK

Model 3060 (stick meter), models, 3400, 3410,
3420, 3100, 3070, 3050 PHM3 PHM4,
PHM5, PHM9 and PHM11 (potable water)
Model 3010 (low-cost) Model 3020
(microprocessor) Model 3030
(microcomputer)

Jenway Ltd
Gransmore Green
Felsted
Dunmow
Essex CM6 3LB
UK

SA 720 (basic), Model 611 (small samples), SA
520 (basic), SA 230 and SA 210
Orion Research UK
Freshfield House
Lewes Road
Forest Row
East Sussex PH18 5ES
UK

Orion Research Incorporated
Laboratory Products Group
The Schraft Centre
529 Main Street
Boston MA 02129
USA

Electrodes only
Ingold Electrodes Inc
261 Ballardvale Street
Wilmington MA 01887
USA

PT 105 and PT 110
Wilkinson & Simpson Ltd
Palintest House
Kingway
Team Valley
Gateshead
Tyne and Wear NE11 0NS
UK

Section 5.2
Temperature measurement

Model 1103
Environmental and Industrial
Measurements E/M (Northern) Ltd
132 Buxton Road
Heavily
Stockport
Cheshire SK2 9PL
UK

Sigma 4
EDT Ltd
Lorne Road
Dover
Kent CT16 2DR
UK

2060, 2000 series
Jenway Ltd
Gransmore Green
Felsted
Dunmow
Essex CM6 3LB
UK

Section 5.3
Electrical conductivity
meters

Beta 81, Beta 82, Beta 800, Sigma 3
EDT Ltd
Lorne Road
Dover
Kent CT16 2DR
UK

SL-1
Orion Research UK
Freshfield House
Lewes Road
Forest Row
East Sussex RH18 5ES
UK

Orion Research Incorporation
Laboratory Products Group
The Schraft Centre
529 Main Street
Boston MA 02129
USA

PCM1, 3410, 3420, 4060, 4070, 4010, 4020
Jenway Ltd
Gransmore Green
Felsted
Dunmow
Essex CH6 3LB
UK

EM2, EM3, EM4 and EM5
Elga Ltd
Lane End
High Wycombe
Buckinghamshire HP14 3JH
UK

PT115
Wilkinson & Simpson Ltd
Palintest House,
Kingsway
Team Valley
Gateshead
Tyne and Wear NE11 0NS
UK

Section 5.4
Dissolved oxygen meters

PT 125
  Wilkinson & Simpson Ltd
  Palintest House
  Kingway
  Team Valley
  Gateshead
  Tyne and Wear NE11 0NS
  UK

ECO 291
  EDT Ltd
  EDT Research
  14 Trading Estate Road
  London NW10 7LU
  UK

9010, 3410 with DO2 probe, POM 2, 9060, 9070
  Jenway Ltd
  Granmore Green
  Felsted
  Dunmow
  Essex CH6 3LB
  UK

SL.9
  Orion Research UK
  Freshfield House
  Lewes Road
  Forest Row
  East Sussex RH18 5ES
  UK

Section 5.5
Colour

CL 4000 colour measurement systems
  Trivector International Ltd
  Sunderland Road
  Sandy
  Bedfordshire SG19 1RB
  UK

  Trivector Inc
  PO Box 2322
  Westchester
  PA 19380
  USA

Section 5.6
Total elements

Total halide (DX 80B), also total organic halogen (DX 20A), total sulphur and chlorine (MCIS 130/120) total nitrogen, (DN–100), total organic carbon DC-80, DC-180, DC-90, DC-85A, DC-88 and DC-54
  Dohrmann Instruments
  Rosemount Analytical Division
  3240 Scott Boulevard
  Santa Clara 90062
  CA
  USA

  Sartec Ltd
  Bourne Industrial Estate
  Wrotham Road
  Borough Green
  Sevenoaks
  Kent TN15 8DG
  UK

Total sulphur, TS 02 and TN 02 and 702X
  Mitsubishi Chemical Industries Ltd
  Instruments Dept
  Mitsubishi Building
  5–2 Marunouchi-2-Chrome
  Chyoda-ku
  Tokyo 100
  Japan

Total sulphur chlorine (TOX–10), total organic halogen (TSX-10), total nitrogen (TN-05, DN-10, DN-100 and N-100)
  EDT Analytical Ltd
  14 Trading Estate Road
  London NW10 7LU
  UK

Total nitrogen
Kjeltex System 1 with kjeltec system 1026 distilling unit also 1030 analyser also system 6120 kjeldahl digestor also system 12/40 kjeldahl digestor
  Tecator AB
  Box 70
  S263-01
  Honagas
  Sweden

System for chemical oxygen demand
  Tecator Ltd
  Cooper Road
  Thornbury
  Bristol BS12 2UP
  UK

Carbon, hydrogen, nitrogen
2400 CHN (combustion gas chromatography)
Perkin-Elmer Ltd
Post Office Lane
Beaconsfield
Buckinghamshire HP9 1QA
UK

Perkin-Elmer Corporation
Analytical Instruments Division
761 Main Avenue
Norwalk CT 06859 – 0012
USA

Nitrogen, carbon and sulphur NA-1500
Carlo Erba Instruments
Strada Rivoltana
20090 Rodano
Milan
Italy

Fisons Instrument
Sussex Manor Park
Gatwick Road
Crawley
West Sussex RH10 2QQ
UK

Total organic carbon – TOC-500
Shimadzu Corporation
International Marketing Division
Shinjuku Mitsui Building
1,1, Nishi Shinjuku 2 Chrome
Shinjuku ku
Tokyo 163
Japan

Dyson Instruments Ltd
Hetton Lyons Industrial Estate
Hetton
Houghton le Spring
Tyne and Wear DH5 0RH
UK

Total organic carbon
Model 700
O I Corporation
Graham Road at Wellborn Road
PO Box 2980
College Station
Texas 77841 2980
USA

Centronic Sales Ltd
Centronic House
King Henry's Drive
New Addington
Croydon
Surrey CR9 0B9
UK

Acid digestion systems for wet digestions
System 6/20
System 12/40
Tecator AB
Box 70
S-203-01
Hanagas
Sweden

Tecator Ltd
Cooper Road
Thornbury
Bristol BS12 2UP
UK

System 500
Skalar BV
PO Box 3237
4800 DE Breda
Netherlands

Section 6.1
Gas chromatography

There are numerous suppliers of gas
chromatography equipment, a selection of
which are given below
Carlo Erba Instruments
Strada Rivoltana
20090 Rodano
Milan
Italy

Models 8100, 8200, 8400, 8500 and 8700
sigma 2000 range
Perkin-Elmer Corporation
Analytical Instruments
761 Main Avenue
Norwalk
CT 06859-0012
USA

Perkin-Elmer Ltd
Post Office Lane
Beaconsfield
Buckinghamshire HP9 1QA
UK

GC 14A, GC 15A, GC 16A, GC 8A
Shimadzu Corporation Division
International Marketing
Shinjuku Mutsui Building
1-1 Nishi Shinjuku 2 Chrome
Shinjuku ku
Tokyo 163
Japan

Dyson Instruments Ltd
Hetton Lyons Industrial Estate

Hetton
Houghton le Spring
Tyne and Wear DH5 3RH
UK

Micromat HRGC 412
 Nordion Instruments Co Ltd
 PO Box 1
 SF 003171 Helsinki
 Finland

Silchromat 1-4 and Silchromat 2-8
 Siemens AG
 Instrumentation and Control Division
 E687
 Postfach 211262
 D-7500 Karlsruhe 21
 Germany

 Siemens Ltd
 VE6
 Siemens House
 Eaton Bank
 Congleton
 Cheshire CW12 1PH
 UK

Section 6.2
Headspace samplers

HS 101, HS 100 and HS 6
 Perkin-Elmer Corporation
 Analytical Instruments Division
 Main Avenue MS-12
 Norwalk CT 06856
 USA

H SS – 2A
 Shimadzu Corporation
 International Marketing Division
 Shinjuku Mitsui Building
 1-1 Nishi Shinjuku 2 Chrome
 Shinjuku ku
 Tokyo 163
 Japan

 Dyson Instruments Ltd
 Hetton Lyons Industrial Estate
 Hetton
 Houghton le Spring
 Tyne and Wear DH5 3RH
 UK

Headspace 6
 Siemens, AG
 Instrumentation and Control Division
 E687

Postfach 21 1262
D-7500 Karlsruhe 21
Germany

Siemens Ltd
VE6 Siemens House
Eaton Bank
Congleton
Cheshire CW12 1PH
UK

Section 6.3
Purge and trap
concentrators

4460A
 OIC Corporation
 Graham Road at Wellborn Road
 PO Box 2980
 College Station
 Texas 77841 2980
 USA

 Eden Scientific
 1 Beechrow
 Ham Common
 Richmond
 Surrey TW10 5HE
 UK

LSC 2000 (concentrator) ALS 2016, 2032,
(discrete automatic samplers) automatic
sample heater ALS 2050 (vial sampling
system)
 Tekmar
 10 Knollcrest Drive
 PO Box 371556
 Cincinnati
 Ohio 45222-1856
 USA

Section 6.4
High-performance liquid
chromatography

2000 series, 2500 series, 5000 series, 5500
series, 9060 diode array detector, 9060 LC
autosampler
 Varian
 220 Humboldt Court
 Sunnyvale
 California 94069
 USA

Varian Associates Ltd
28 Manor Road
Walton on Thames
Surrey KT12 2QF
UK

Series 10 chromatography LC-95 variable
wavelength UV/visible detector, LC-90
variable wavelength UV detector, LC-135 and
L-235 diode array detectors, LC 1-100
computing integrator, 1SS-100 intelligent
sampling system, Series 410 LC pump
  Perkin-Elmer Corporation
  Analytical Instruments Division
  761 Main Avenue
  Norwalk CT 06859
  USA

  Perkin-Elmer Ltd
  Post Office Lane
  Beaconsfield
  Buckinghamshire HP9 1QA
  UK

System 400, comprising 420 and 414 pumps,
460 autosampler, 430 and 432 detectors, 450
data system, 480 column oven, 425 gradient
former, Anacomp 220 data management, 306
autosampler, MSI 66 autosampler, 740 LC
variable wavelength detector, 720 LC digital
variable wavelength detector, 735 LCC
variable wavelength detector
  Kontron Instruments
  Blackmore Lane
  Croxley Centre
  Watford
  Hertfordshire WD1 8XQ
  UK

Series 4500i
  Dionex Corporation
  PO Box 3063
  Sunnyvale
  California
  USA

  Dionex (UK) Ltd
  Albany Park
  Camberley
  Surrey GU15 2PL
  UK

2350 pump, 2360 gradient programmer, 2351
gradient controller V4, UA5 and 228
wavelength detectors, FL-2 fluorescence
detector, Chem Research data management/
gradient control system, 015A recorder,
autoinjector, Foxy, Retriever II and Cygnet
fraction collectors, Peak collection instrument
  Isco
  4700 Superior

Lincoln
NE 67504
USA

2150 pump, 2249 gradient pump, 2152
controller, 2157 autosampler, 2154 injector,
2510 UV detector, 2151 variable wavelength
detector, 2140 rapid spectral detector, 2142
refractive index detector, 2143
electrochemical detector, 2221 integrator,
2145 data system, 2210 and 2240 recorders,
2134, 2133, 2135, 2134, 2131 column
  Pharmacia LKB
  Bjorkgaten 30
  75182 Uppsala
  Sweden

  Pharmacia Ltd
  Pharmacea LKB Technology Division
  Midsummer Boulevard
  Central Milton Keynes
  Buckinghamshire MK9 3HP
  UK

LC 6A system comprising SCL-6A controller,
SPD-6A and SPD-6AV spectrophotometric
detectors, CTO-6A column oven, LC-6A
pump, SIL-6A autoinjector

LC-8A preparative HPLC comprising LC-8A
pump, FCV-110AL reservoir switching valve,
FCV-130 AL valve/pump box, FCV-120AL
recycle valve, SCL-8A system controller,
SPD-6A UV detector, SPD, 6AV UV-visible
detector, SIL-8A autoinjector, 7125 manual
injector, FCV 100B fraction collector, C-R4A
data processor, PC-11L 3-pump interface,
PC-30L pump interface, PC-24L interface for
reservoir switching valve FCV 110 AL and
recycle valve FCV 120 AL, PC-16N interface
for fraction collector FCV 100B, PC-14N
interface for data processor C-R4A, LC-7A
bicompatible system comprising LC-7A
pump, LC-7A gradient system, 7125/T
sample injector, SPD-7A UV detector, SPD
7AV UV/visible detector
  Shimadzu Corporation
  International Marketing Division
  Shinjuku Mitsui Buildings
  1-1 Nishi Shinjuku 2 Chrome
  Shinjuku ku
  Tokyo 163
  Japan

  Dyson Instruments Ltd
  Hetton Lyons Industrial Estate
  Hetton
  Houghton le Spring
  Tyne and Wear DH5 3RH
  UK

HP 1050 Series comprising programmable variable-wavelength detector, multiple-wavelength detector, pumping system, autosampler
Hewlett Packard
PO Box 10301
Palo Alto
California 94303 0890
USA

HPLC columns only
HPLC Technology
Wellington House
Waterloo St West
Macclesfield
Cheshire SK11 6PJ
UK

Spectroflow 400 system
HPLC pump only
Kratos Analytical Instruments
170 Williams Drive
Ramsey
New Jersey NJ 07446
USA

Chromo-A-Scope comprising UV visible detector and data processing system only
Barspec Ltd
PO Box 560
Rehovot 76103
Israel

Barspec Ltd
PO Box 430
Mansfield
MA 02048
USA

Roth Scientific Co Ltd
Alpha House
Alexandra Road
Farnborough
Hampshire GU14 6BU
UK

Hichrome Ltd
6 Chiltern Enterprise Centre
Station Road
Theale
Reading
Berkshire RG7 4AA
UK

Series 100 comprising CE 1100 pump, CE 1200 variable wavelength monitor, CE 1300 gradient programmer, CE 1400 refractive index detector, CE 1500 electrochemical detector, CE 1700 computing integrator, CE 1710 recorder and 1720 recorder, CE 1800 sample injector, 1200, 2000 column monitoring panel and sample valve.

Cecil Instruments Ltd
Milton Technical Centre
Cambridge CB4 4AZ
UK

Model 5100 A Coulochem electrochemical detector only
ESA Inc
45 Wiggins Avenue
Bedford MA 01730
USA

Severn Analytical
30 Brunswick Road
Gloucestershire GL1 1JJ
UK

Model RR/066 351 and 352 pumps: models 750/16 variable-wavelength UV monitor detector 750/11 variable filter UV detector, MPD 880S multiwave plasma detector, 750/14 mass detector 750/350/06 electrochemical detector refractive index detector; HPLC columns: column heaters, autosamplers, pre-columns derivatization systems, solvent degassers, preparative HPLC systems
Applied Chromatography Systems Ltd
The Arsenal
Heapy Street
Macclesfield
Cheshire SK11 7JB
UK

LCA 15 system; LCA 16 system
EDT Ltd
EDT Research
14 Trading Estate Road
London NW10 7LU
UK

Aspec Automatic sample preparation system; Asted automated sequence trace diazylate enricher; 231/401 HPLC autosampling injector: 232/401 automatic sample processor and injector:
Gibson Medical Electronics (France) SA
72 Rue Gambetta
BP45 F-95400
Villiers le Bel
France

Gilson Medical Electronics Incorporated
Box 27
300 010 Beltine Highway
Middleton WC 53562
USA

Isoflo HPLC radioactivity monitor, iso mix interface between HPLC column and Isoflo HPLC radioactivity monitor:
Nuclear Enterprises Ltd
Bath Road

Beenham
Reading
Berkshire RG7 5PR
UK

Advanced automated sample processor:
  Varian AG
  Steinhauserstrasse CH 6300
  Zug
  Switzerland

  Varian Associates
  28 Manor Road
  Walton on Thames
  Surrey KT12 2QF
  UK

High-performance liquid chromatography – mass spectrometry

HP 5988 A Mass selective detector; HP 4987
A Mass selective detector:
  Hewlett Packard
  PO Box 10301
  Palo Alto
  California 94303-0890
  USA

Section 6.5
Supercritical fluid chromatography

501 SFC, 602 SFC, 622 SFC/GC
  Lee Scientific
  4426 SO Century Drive
  Salt Lake City
  Utah 84123-2513
  USA

  Dionex (UK) Ltd
  Selmoor Road
  Farnborough
  Hampshire GU14 7QN
  UK

Supercritical fluid chromatography pressure
gradient solvent delivery system:
  Pierce Chemicals
  Life Science Laboratories Ltd
  Sedgewick Road
  Luton
  Bedfordshire LU4 9DT
  UK

CCS 5000
  Severn Analytical Ltd
  Unit 2B
  St Frances' Way
  Shefford Industrial Estate
  Shefford
  Bedfordshire SG17 5DZ
  UK

CW 14 Detector
  Valco Instruments Co Ltd
  PO Box 55603
  Houston
  Texas 77255
  USA

  Valco Europe
  Untertannberg 7 CH-6214
  Schenkon
  Switzerland

Section 6.6
Gas chromatography – mass spectrometry

SSQ 70 Series Single stage gas
chromatography quadruple mass spectrometer
TSQ-70 series gas chromatograph triple-stage
quadrupole mass spectrometer; MAT-90 gas
chromatography high-resolution mass
spectrometer; H SQ-30 hybrid mass
spectrometer – mass spectrometer; series 700
ion-trap detector; Incas-50 gas-
chromatography quadruple mass
spectrometer; ChemMaster workstation, 1020
quadruple mass spectrometer, OWA-20/30B
organics in water gas chromatograph–mass
spectrometer (see above)
  Finnigan MAT
  355 River Oaks Parkway
  San Jose
  California
  CA 9513-1991
  USA

  Finnigan MAT Ltd
  Paradise
  Hemel Hempstead
  Hertfordshire HP2 4TG
  UK

Ion-trap detector
  Perkin-Elmer Corporation
  Analytical Instruments Division
  761 Main Avenue

Norwalk
Connecticut 06859-0012
USA

Perkin-Elmer Ltd
Post Office Lane
Beaconsfield
Buckinghamshire HP9 1QA
UK

GCMS-QP 2000 gas chromatograph mass
spectrometer, GCMS-Q 1000 and GC-QP
1000A gas chromatograph mass spectrometer;
Mispack 200 GCMS QP series MS data system
  Shimadzu Corporation
  International Marketing Division
  Shinjuku
  Mitsui Building
  1-1 Nishi Shinjuku-2-Chrome
  Shinjuku-ku
  Tokyo 163
  Japan

  Dyson Instruments Ltd
  Hetton Lyons Industrial Estate
  Hetton
  Houghton le Spring
  Tyne & Wear DH5 3RH
  UK

Section 6.7
Fourier transform infrared
spectroscopy

FTIR
  Spectratech Europe Ltd
  Genesis Centre
  Science Park South
  Birchwood
  Warrington
  Cheshire WA3 7BH
  UK

FTIR
  Mattson Instruments Inc
  1001 Fourier Court
  Madison
  Wisconsin W183717
  USA

  Mattson Instruments Ltd
  Linford Forum
  Rockingham Drive
  Linford Wood
  Milton Keynes
  Buckinghamshire MK14 6LY
  UK

FTIR Model 1800, 1700 series
  Perkin-Elmer Corporation
  Analytical Instruments Division
  761 Main Avenue (MS-12)
  Norwalk
  Connecticut 06859
  USA

  Perkin-Elmer Ltd
  Post Office Lane
  Beaconsfield
  Buckinghamshire HP9 1QA
  UK

Digilab TT57
  Bio-Rad
  53/56 Greenhill Crescent
  Watford Business Park
  Watford
  Hertfordshire WD1 8QS
  UK

PV 9800
  Phillips Electronic Instruments
  85 McKee Drive
  Mohway
  NJ 07430
  USA

  Phillips Nederland BV
  Asd Analysetechnickin VB 3
  Postbus 90050
  5600 PB Eindhoven
  Netherlands

Section 6.8
Amino acid analysers

Model 4150 alpha; model 4151 alpha plus
  LKB Biochrom Ltd
  Science Park
  Cambridge CB4 4BH
  UK

  Pharmacia LKB Biotechnology
  AB, 75182 Uppsala
  Sweden

Bio-LC amino acid system
  Dionex UK Ltd
  Selmoor Road
  Farnborough
  Hampshire GU15 2PL
  UK

  Dionex Corporation
  PO Box 3063
  Sunnyvale
  California
  USA

Section 6.9
Luminescence instruments and spectrofluorimeters

Luminescence instrument LS-3B;
luminescence instrument LS-5B; Accessories:
low flow cell, cell holders, bioluminescence
spectroscopy, fluorescence spectroscopy,
recorders/printers, low-temperature
luminescence, fluorescence plate reader,
polarization accessory, microfilm fluorimeter
LS-2B
   Perkin-Elmer Ltd
   Analytical Instruments Division
   761 Main Avenue
   Norwalk
   Connecticut 068-59-0012
   USA

   Perkin-Elmer Ltd
   Post Office Lane
   Beaconsfield
   Buckinghamshire HP9 1QA
   UK

SFM-25 spectrofluorimeter
   Kontron Instruments
   Kontron AG
   Bernerstrasse Sud 169
   8010 Zurich
   Switzerland

Chemi and bioluminescence; lumicon,
luminescence instrument
   Hamilton Co
   PO Box 10030
   Reno
   Nevada 89520
   USA

   Hamilton Bonaduz AG
   PO Box 26
   CH-7402 Bonaduz
   Switzerland

Section 6.10
NMR Spectroscopy

Gemini Superconducting Fourier transform
NMR systems, VXR series 5
   Varian Instruments
   Sugar Lane
   Texas
   USA

NMR imaging spectrometer systems
   Vis
   1120 Auburn Road
   Fremont
   California 945 38
   USA

Section 7.1
Biology laboratory equipment

Autoclaves: Imperial, Prince, Sovereign
   Denley Instruments Ltd
   Notts lane
   Billingshurst
   West Sussex RH14 9EY
   UK

AUX-750 (front loading) AUX-755 (top
loading)
   Gallenkamp Ltd
   Belton Road West
   Loughborough
   Leicestershire LE11 0TR
   UK

PC/LAC/V/100 (front loading) PC/LAC/V/150
(top loading)
   The Northern Media Supply Co
   Sainsbury Way
   Hessle
   North Humberside HU13 9NX
   UK

Microcentrifuges: BM 402 (unrefrigerated),
BR 400 (refrigerated)
   Denley Instruments Ltd
   Notts Lane
   Billingshurst
   West Sussex RH14 9EY
   UK

Refrigerators and freezers: wide range
available: refrigerators 2.1–7.1 ft^3, freezers
1.7–5.8 ft^3, freezers 540u 15 ft^3, 1740u 17 ft^3,
1035u 9.4 ft^3, 835c 8 ft^3, 180L 61 ft^3, 309 L
11 ft^3, 360L 10.5 ft^3, 487L 17.3 ft^3, 701L
25.0 ft^3
   The Northern Media Supply Co
   Sainsbury Way
   Hessle
   North Humberside HU13 9NX
   UK

   Gallenkamp Ltd
   Belton Road West

Loughborough
Leicestershire LE11 0TR
UK

Refrigerators cold stores: KI 229 (2 - 2 tray);
K2 346 (3 - 5 tray); K 2.5 433 (4 - 6 tray); K3
463 (4 - 6 tray); K4 581 (5 - 7 tray)
   Denley Instruments Ltd
   Notts Lane
   Billingshurst
   West Sussex RH14 9EY
   UK

Horizontal laminar air flow cabinet
   Gallenkamp Ltd
   Belton Road West
   Loughborough
   Leicestershire LE11 0TR
   UK

Controlled Environment rooms:
   Vindon Scientific Ltd
   Ceramyl Works
   Diggle
   Oldham
   Lancashire OL3 5JY
   UK

Incubators: orbital incubator (refrigerated and
unrefrigerated), water-jacketed carbon
dioxide incubator (170 l), dry-wall carbon
dioxide incubator, Plus series culturing
incubators; Prime series high-performance
incubators, economy incubators (56–210 l),
cooled incubators (180–400 l); Wellwarm 1
shaker incubator (l = capacity in litres)
   Gallenkamp Ltd
   Belton Road West
   Loughborough
   Leicestershire LE11 0TR
   UK

Wellwarm 1 shaker incubator
   Denley Instruments Ltd
   Notts Lane
   Billingshurst
   West Sussex RH14 9EY
   UK

Maxi range: anhydric incubators: MAX 47
capacity 47 l, MAX 68 (68 l), MAX 118
(118 l), MAX 156 (156 l), MAX 185 (185 l).
Large capacity: 9V (capacity 250 l), 12V
(335 l), 15V (435 l), 18V (530 l), 8H (225 l),
12H (335 l), 15H (425 l), 18H (510 l), 22H
(630 l), 24H (675 l)
   The Northern Media Supply Co
   Sainsbury Way
   Hessle
   North Humberside HU13 9NX
   UK

Model 2286 (80 l) (non-refrigerated), 2386
(157 l), 2486 (236 l), 2589 (330 l); Model 513
(210 l) (refrigerated), 513S (330 l), 516 (570 l)
   Vindon Scientific Ltd
   Ceramyl Works
   Diggle
   Oldham
   Lancashire OL3 5JY
   UK

Laboratory ovens: MIN 696 (capacity 5 l),
MIN 1 (16 l), MIN 4 (28 l), MID 1 (33 l), MID
2 (53 l), MID 3 (75 l), MID 4 (113 l), MID 5
(120 l), MID 6 (185 l), MOBS (47 l), MOBS
(68 l), MOBS (118 l), MOBS (156 l), MOBS
(185 l), MAX 47 (47 l), MAX 68 (68 l), MAX
118 (118 l), MAX 156 (156 l), MAX 185
(185 l), 9V (256 l), 12V (335 l), 15V (435 l),
18V (530 l), 8 H (225 l), 12H (335 l), 15 l
(425 l), 18 l (510 l), 22H (630 l), 24H (675 l),
26H (720 l), 30H (850 l), 35H (1000 l); Model
112 (80 l), 113 (157 l), 114 (236 l), 116 (750 l)
(Models 112–16 all operate at up to 300°C)
   Vindon Scientific Ltd
   Ceramyl Works
   Diggle
   Oldham
   Lancashire OL3 5JY
   UK

Microscopes: Olympus CH 2 series
   The Northern Media Supply Co
   Sainsbury Way
   Hessle
   North Humberside HU13 9NX
   UK

Pipettes: Powerpette (0.1–100 ml),
Powerpette junior (0.1–100 ml), Sealpette
fixed economy (100–1000 µl), Sealpette
variable (2–100 µl), Sealpette variable
(2–1000 µl), Sealpette universal
(100–1000 µl), Sealpette clinical (0.5–5000 µl)
   Jencons Scientific Ltd
   Cherrycourt Way Industrial Estate
   Stanbridge Road
   Leighton Buzzard
   Bedfordshire LU7 8UA
   UK

700 series (5–500 µl), 700 series Rheodyne
(25–500 µl), 800 series (5–250 µl), 900 series
(0.05–10 µl), 1800 series (10–250 µl), 1000
series gastight (10–5000 µl), 7000 series (0.5–
25 µl)
Microlab P electrically operated micropipette
(0.5–5000 µl)
   V A Howe & Co Ltd
   12–14 St Ann's Crescent
   London SW18 2LS
   UK

Tecator Ltd
Tecator House
Cooper Road
Thornbury
Bristol BS12 2UW
UK

Hamilton Bonaduz AG
PO Box 26
CH-7402 Bonaduz
Switzerland

Electric pipettes, Microelectrapipette
(1–10 000 µl)
Arnold R. Horwell Ltd
73 Maygrove Road
West Hampstead
London NW6 2BP
UK

Reagent dispensing systems: Zipette Mark IV
(0.5–100 µl), Electronic Zipette (1–10 ml) and
(3–30 ml), Digitrate Titrator (0–25 ml) and
(0–50 ml)

Perifill 1Q2000 (0.5–1000 ml)
Jencons Scientific Ltd
Cherrycourt Way Industrial Estate
Stanbridge Road
Leighton Buzzard
Bedfordshire LU7 8UA
UK

Precision liquid dispenser II (1–20 µl)
electrically (pneumatically) operated,
Microlab 1000 programmable diluter/
dispenser (1–2500 ml) (electrically operated).
Microlab P programmable pipette (electrically
operated), Microlab 400 modular diluters and
dispensers (electrically operated), Microlab M
programmable diluter/dispenser (50 µl–50 ml)
(electrically operated), Microlab MT
microlitre plate dilutor/dispenser (5–1000 µl)
(electrically operated)
Hamilton Bonaduz AG
PO Box 26
CH-7402 Bonaduz
Switzerland

V A Howe & Co Ltd
12–14 St Ann's Crescent
London SW18 2LS
UK

Reagent dispensing systems (1–10 000 µl)
(electrically operated)
Berghof GmbH
Harretstrasse 1 D7412
Eningen UA
Germany

Ice flakers: AF10 (70 kg ice per day), AF20
(130 kg ice per day), AF30 (180 kg ice per day)
Denley Instruments Ltd
Notts Lane
Billingshurst
West Sussex RH14 9EY
UK

Reciprocal mixers: RM251 Reciprocal mixer,
Reciproload orbital mixers
(28 × 50 ml flasks)
Denley Instruments Ltd
Notts Lane
Billingshurst
West Sussex RH14 9EY
UK

Tube racks
Denley Instruments Ltd
Notts Lane
Billingshurst
West Sussex RH14 9EY
UK

Modular storage systems: Denleyrack
modular storage systems, Denleystore drawer
modules and microscope slide store systems,
specimen collection and transport systems
Denley Instruments Ltd
Notts Lane
Billingshurst
West Sussex RH14 9EY
UK

Fluorimeters: Microfilm fluorimeter LS-2B
Perkin-Elmer Corporation
Analytical Instruments Division
761 Main Avenue
Norwalk
Connecticut 06859-0012
USA

Perkin-Elmer Ltd
Post Office Lane
Beaconsfield
Buckinghamshire HP9 1QA
UK

Luminometers: Unicam Chemi and
bioluminescence luminometer
Hamilton Co
PO Box 10030
Reno
Nevada 89520
USA

Hamilton Bonaduz AG
PO Box 26
CH-7042 Bonaduz
Switzerland

Visible spectrophotometers: EPOS analyser
5060, data terminal 640 selective
spectrophotometric analysis system:
Eppendorf Geratebau
Netheler and Hinz GmbH
PO Box 650670
D-2000 Hamburg 65
Germany

British Drug Houses Ltd
Broome Road
Poole
Dorset BH12 4NN
UK

Carbon dioxide analyser: 965 D carbon
dioxide analyser
Denley Instruments Ltd
Notts Lane
Billingshurst
West Sussex RH14 9EY
UK

Biosensor electrodes
Electrochemical Sensors plc
Cambridge Life Sciences
Science Park
Milton Road
Cambridge
Cambridgeshire CB4 4GN
UK

Bacteria and virus removal: Anopore range
Alltech Association
Applied Science Ltd
Kellet Road Industrial Estate
Carnforth
Lancashire LA5 9XP
UK

Millipore Intertech
PO Box 255
Bedford MA 01730
USA

Millipore (UK) Ltd
11–15 Peterborough Rd
Harrow
Middlesex HA1 2YH
UK

Section 7.3.1
Partially automated
immunoassay systems

Rotary platers for petri dishes: RP 453, RP
454; Autospreader for petri dishes: A 450;
Multipoint spreader for petri dishes: A400;
Single-reagent dispenser for microplates:

Wellfill 3; Multi-reagent dispenser for
microplates: Wellfill 4; Single-reagent plate-
stacking dispenser for microplates: Wellfill 5;
Microplate innoculation: Wellrepp 2
automatic replicator; Shaker for microplates:
Wellmix X1, Wellmix X2, Wellmix X3,
Wellmix X4
Denley Instruments Ltd
Notts Lane
Billingshurst
West Sussex RH14 9EY
UK

Microlab AT automated pipetter diluter/
distributor for microplates
Hamilton Bonaduz AG
PO Box 26
CH-7402 Bonaduz
Switzerland

Fluorimeters: Microfilm fluorimeter LS-2B
Perkin-Elmer Corporation
Analytical Instruments Division
761 Main Avenue
Norwalk
Connecticut 06859-0012
USA

Perkin-Elmer Ltd
Post Office Lane
Beaconsfield
Buckinghamshire HP9 1QA
UK

Unicam chemi- and bioluminescence
luminometer
Hamilton Co
PO Box 10030
Reno
Nevada 59520
USA

Hamilton Bonaduz AG
PO Box 26
CH-7042 Bonaduz
Switzerland

Lambda reader automated microplate reader
Perkin-Elmer Corporation
Analytical Instruments Division
761 Main Avenue
Norwalk
Connecticut 06859-0012
USA

Perkin-Elmer Ltd
Post Office Lane
Beaconsfield
Buckinghamshire HP9 1QA
UK

Ependorf Geratebau
Netherler and Hius GmbH

PO Box 650670
D-2000 Hamburg 65
Germany

MDA 312 Multidetector radio-immunoassay
analyser
    Kontron AG
    Bernerstrasse Sud 169
    8010 Zurich
    Switzerland

Cobra-one Auto Gamma 5012 or 5013 crystal
plus bench top radio-immunoassay system
    Packard Instrument Co
    2200 Warrenville Road
    Downers Grove
    Illinois 60515
    USA

    Canberra Packard International SA
    Peuggesstrasse 3
    CH-8038 Zurich
    Switzerland

Section 7.3.2
Fully automated
immunoassay
workstations

Zymark Ryobotic Laboratory Automation
system for immunoassays
    Zymark Corp
    Zymark Centre
    Hopkinton 1
    MA 01748
    USA

    Zymark Ltd
    The Genesis Centre
    Science Park South
    Birchwood
    Warrington
    Cheshire WA3 7BH
    UK

Macrogroupamatic
    Kontron Instruments SA
    2 Avenue du Manet
    781 80 Montigny-le-Britonneux
    France

    Kontron AG
    Bernerstrasse Sud 169
    8101 Zurich
    Switzerland

Biomek 1000 automated laboratory
workstation
    Beckman Instruments Inc
    Spinco Division
    1050 Page Mill Road
    Palo Alto
    California 94304
    USA

    Beckman RIIC Ltd
    Progress Road
    Sands Industrial Estate
    High Wycombe
    Buckinghamshire HP12 4JL
    UK

Section 7.4
Laboratory homogenizers

Planetary micromill Pulverisette 7, vibration
micro pulverizer, Pulverisette 0, rotary speed
mill, Pulverisette 14
    Fritsch GmbH
    Laborgeraetebau
    Industriesstrasse 8
    D-6580 Idar Oberstein
    Germany

Laboratory comminuters: vibrating cup mill,
Pulverisette 9, laboratory disk mill,
Pulverisette 13, mortal-grinders, Pulverisette
2, centrifugal mill, Pulverisette 6, planetary
mill, Pulverisette 5

Laboratory sieving devices: vibratory sieve
shaken for micro-precision sieving, Analysette
3, rotary sieve shaker, Analysette 18
    Christison Scientific Equipment Ltd
    Albany Road
    Gateshead
    Tyne & Wear NH8 3AT
    UK

Particle size distribution measurement:
(a) Sedimentation in gravitational field,
Analysette 20;
(b) Laser diffraction, Analysette 22
(c) Sedimentation in centrifugal field,
Analysette 21
    Fritsch GmbH
    Laborgeraetbau Industriestrasse 8
    D 6580 Idar Oberstein
    Germany

    Christison Scientific Equipment Ltd
    Albany Road
    Gateshead
    Tyne & Wear NH8 3AT
    UK

Electrical zone sensing Model ZM Coulter
Multisizer
  Coulter Electronics Inc
  Hialeah
  Florida
  USA

  Coulter Electronics Ltd
  Northwell Drive
  Luton
  Bedfordshire LU3 3RH
  UK

Acid digestion of solid samples
Pressure dissolution and digestion bombs:
Nos 4745, 4749, 4744, 4746, 4748, Microwave
and digestion bombs: Nos 4781, 4782,
Oxygen combustion bomb: 1108
  Parr Instruments Co
  211 Fiftythird Street
  Moline
  Illinois 61265
  USA

Microwave Oven MD 581D
  CEM Corporation
  PO Box 200
  Matthews NC 28106
  USA

  Oxford Laboratories
  Page House
  164 West Wycombe Road
  High Wycombe
  Buckinghamshire HP12 3AE
  UK

Microdigest 300 and Microdigest 300A
  Prolabo
  23 Rue Pelee 75011
  Paris
  France

BOD Analysers
Manometric BOD analyser, model 212 (6
place), Model 214 (10 place) and
thermostatically controlled BOD cabinets
models 180 (180 l capacity) and model 370
(370 l capacity)
  PHOX Systems Ltd
  Ivel Road
  Shefford
  Bedfordshire SG17 5JU
  UK

BOD tester model 025-601
  Tech-line Instruments
  Tri Campus Park
  PO Box 1236
  Ford du Lac
  Wisconsin 54935
  USA

Foss Electric (UK) Ltd
The Chantry
Bishopthorpe
York YO2 1QF
UK

Sample Processor 100
  Skalar BV
  PO Box 3237
  4800 DE Breda
  Netherlands

Zymate 11 BOD System comprising Z100-X
Zymate 11 system, Z330 pH millivolt
temperature module; Z510 Master laboratory
station, Z820 printer; Z830 Power and event
control station, Z900 General purpose hand,
Z910 precision microlitre syringe hand, DO/
diluent water station, BOD capping station
  Zymark Corporation
  Zymark Centre
  Hopkinton MA 01748
  USA

  Zymark Ltd
  The Genesis Centre
  Science Park South
  Birchwood
  Warrington
  Cheshire WA3 7BH
  UK

Chapter 8
Radioactivity
measurements

Low-level $\alpha\beta\gamma$ counting system 2401 and
2401F (manual system), 2400 and 2400F
(automatic system), series 95 multichannel
analyser, series 10 plus portable multi-channel
analysis system, germanium detectors, well
type, extended range type, coaxial type,
low-energy type, reverse electrode coaxial type
and planar type, lead shield 747 cryostats
7500, 7500 SL, 7600 and 7913 -30 types,
portable cryostats
  Canberra Packard Ltd
  Brook House
  14 Station Road
  Pangbourne
  Berkshire RG8 7DF
  UK

Segmented γScanner (234 U and 239 U) assays in waste. Whole body scanners 2250, 2260, 2270, Abacus 11
Canberra Instruments Ltd
One State Street
Meriden
Connecticut 06450
USA

Series 10 plus portable gamma multi-channel analyser for field use 7404 Quad Alpha, alpha spectroscopy system
Canberra Industries Ltd
45 Gracey Street
Meriden
Connecticut 06450
USA

Gammamatic I/II automatic gamma counting system
Kontron AG
Bernerstrasse Sud 169
8010 Zurich
Switzerland

Betamatic IV-V liquid scintillation counters
Kontron Instruments Ltd
Blackmoor Lane
Croxley Centre
Watford
Hertfordshire WD1 8XQ
UK

Tricarb/LL series, low-level liquid, scintillation analysers, Tricarb 2050 CA, Tricarb 1550, Tricarb 1050, Tricarb 2250, Tricarb 100 (low cost), Tricarb 1500 sample changer, Tricarb 1900 CA computer, Tricarb 2200 CA colour graphics. Auto-gamma 5550/5530 gamma spectrometer, Cobra 5010/5005 autogamma automatic multi-detector gamma counter
Packard Instrument Co
2200 Warrenville Road
Downers Grove
Illinois 60515
USA

Canberra Packard International SA
Renggerstrasse 3
CH-8038 Zurich
Switzerland

Labted scale/rate meter alpha-beta scintillators analyser
Bicron Corporation
12345 Kingsman Road
Newbury
Ohio 44065
USA

Bicron Corporation
PO Box 271
2410 A-G Bodegraven
Netherlands

92X Spectrum master gamma spectroscopy workstation
EG&G Ortec
100 Midlands Road
Oak Ridge
Tennessee TN 37831-0895
USA

EG&G Instruments
Division of EG&G Ltd
Bracknell
Berkshire
UK

Lead castles, encapsulated liquid scintillators, flow cells, crystal scintillators, Na (T1), CgI (Na or T1) Ca F; Manual liquid scintillation counter, (tritium and 14 C)
Nuclear Enterprises Ltd
Bath Road
Beenham
Reading
Berkshire RG7 5PR
UK

Integral sodium iodide detectors for gamma spectrometry, Planchette alpha-beta counting and analysis systems, PSR 8 Portable scaler, ST7 Scaler timer, SR8 Scaler ratemeter
NE Technology Ltd
Bankhead
Medway
Sighthill
Edinburgh EH11 4BY
Scotland

Radon measuring systems: 20 MCA Radon Counting system (gamma)
Canberra Industries Inc
One State Street
Meriden
Connecticut 06450
USA

Ortec Airguard Radon Monitoring system (gamma)
EG&G Ortec
100 Midlands Road
Oak Ridge
Tennessee
TN 37831-0895
USA

EG&G Industries Ltd
Division of EG&G Ltd
Bracknell

Berkshire
UK

Picorad Radon Analysis system with Tricarb
1900 CA, Tricarb 1500 or Tricarb 2200 CA
liquid scintillation analysers
    Packard Instruments Co
    2200 Warrenville Road
    Downers Grove
    Illinois 60515
    USA

Tritium analysis – radioactive gas monitor RG
M1/1
    Nuclear Enterprises
    Bath Road
    Beenham
    Reading
    Berkshire RG7 5PR
    UK

Personal radiation monitoring probes
Alpha probes AP2, AP3 and AP4, alpha beta
probes DP3 and DP2, beta probes BP5, BD5
and BP4, high-energy gamma probes BP3/4A,
beta X-ray probe BP1/4A, gamma X-ray
probes GP6, GP7 and GP9/4A

Personal gamma monitoring system PPD1/Si,
Obex contamination monitor (125 I, beta and
X-rays), portable beta gamma dosimeter
PPDM1, gamma inspection monitor PDR4 –
SV, Portable gamma survey monitor PDR3 –
SV and PDR2 SV, low-level gamma radiation
monitor PDR ISV, portable beta–
gamma contamination meter, PCM 5,
radiation gamma doserate meter RDM 1,
ratemeter RH6 for alpha, beta and gamma
probes, portable scaler ratemeter PS R8
    Nuclear Enterprises Ltd
    Bath Road
    Beenham
    Reading
    Berkshire RG7 5PR
    UK

'Analyst' Portable analyser, 'Microanalyst' μR
gamma, ion chamber, survey meter, R7050
and RG050 beta, gamma, X-ray ion chamber
survey meters
    Bicron Corporation
    12345 Kinsman Road
    Newbury
    Ohio 44065
    USA

    Bicron Corporation
    PO Box 271
    2410 AG Bodegraven
    Netherlands

Chapter 9
On-site measuring instruments

Test kits: Merckoquant test strips,
Aquamerck test kits, Aquaquant test kits,
Microquant test kits, Spectroquant analysis
system
    E. Merck
    Postfach 4119
    D-6100 Darmstadt
    Germany

Palintest kits: Visual kits, spectrophotometer
5000 kits:
    Wilkinson & Simpson Ltd
    Palintest House
    Kingway
    Team Valley
    Gateshead
    Tyne and Wear NH11 0NS
    UK

Photometric test kits plus LASA Aqua filter
photometer, LPIW filter photometer, Cadas
100 spectrophotometer
    Dr Bruno Longe GmbH
    Vertriebsbereich Industrieme B gerate
    Willstraterstrasse II
    D-400-Dusseldorf II
    Germany

    Robin Instruments Ltd
    PO Box 93
    Camberley
    Surrey GU15 1DU
    UK

Single-parameter meters: Palintest PT 151 pH
stick meter, Palintest PT 152 and PT 153 total
dissolved solids stick meters, Palintest PT
154, PT 156, PT 157 temperature stick meters
    Palintest Ltd
    Kingway
    Team Valley
    Gateshead
    Tyne & Wear NH11 0NS
    UK

Bibby SMPI pH, mV temperature stick
meter; Bibby SMC1 electrical conductivity
stick meter, Bibby SMO1 dissolved oxygen/
temperature stick meter
    J Bibby Science Products Ltd
    Stone
    Staffordshire ST15 0SA
    UK

Model 21 pH indicator, model 47 battery operated pH recorder, model 42E weatherproof pH indicator, model 42 pH indicator with analogue display, Model 57 battery-operated EC or TDS recorder; series 80 battery-operated flowmeter; model 52E weatherproof EC or TDS recorder; model 67 battery-operated dissolved oxygen recorder; model 62TE weatherproof dissolved oxygen indicator, model 62 digital LCD dissolved oxygen indicator, 3050 probe pH meter, 3061 stick pH meter, 3070 or 3100 probe microprocessor pH/mV meter, 4060 stick electrical conductivity meter, 4070 probe electrical conductivity meter, 9060 stick dissolved oxygen meter, 9070 probe dissolved oxygen/temperature meter

> PHOX Systems Ltd
> Ivel Road
> Shefford
> Bedfordshire SG17 5JU
> UK

> Jenway Ltd
> Gransmore Green
> Felsted
> Dunmow
> Essex CM6 3LB
> UK

Multiparameter meters: Horiba U-7, pH, electrical conductivity, total dissolved solids, temperature and turbidity

> Horiba International Corporation
> 1021 Duryea Avenue
> Irvine
> California 92714
> USA

> Sartec Ltd
> Bourne Industrial Estate
> Wrotham Road
> Borough Green
> Sevenoaks
> Kent TN15 8DG
> UK

JN 3465-00 pH, mV, temperature, conductivity, dissolved oxygen portable electrochemistry analyser

> CP Instruments Co
> PO Box 22
> Bishops Stortford
> Hertfordshire CH23 3DX
> UK

> Jenway Ltd
> Gransmore Green
> Felsted
> Dunmow
> Essex CM6 3LB
> UK

Portable 4–6 channel monitors/dissolved oxygen, temperature, pH and redox, conductivity or total dissolved solids, suspended solids or turbidity, water flow volume or level, nitrate ammonia, organic pollution:

> PHOX Systems Ltd
> Ivel Road
> Shefford
> Bedfordshire SG17 5JU
> UK

Mobile trailer monitoring system

> Jenway Ltd
> Gransmore Green
> Felsted
> Dunmow
> Essex CM6 3LB
> UK

Portable metals analyser: PDV 2000 portable electrochemical analyser

> EDT Ltd
> EDT Research
> 14 Trading Estate Road
> London NW10 7LU
> UK

Section 10.1
On-line process measuring instruments

Single-parameter instruments:
pH: 3080 panel-mounted pH meter with recorder, 3080 HL panel-mounted pH meter with Hi–Lo alarm and relay output, 3090 waterproof panel-mounted pH meter with recorder, 3090 HL panel-mounted pH meter with Hi–Lo alarm and relay output, PHOX 40 pH/redox recorder panel-mounted, PHOX 45 weatherproof panel-mounted pH/redox recorder:

> PHOX Systems Ltd
> Ivel Road
> Shefford
> Bedfordshire SG17 5JU
> UK

> Jenway Ltd
> Gransmore Green
> Felsted
> Dunmow
> Essex CM6 3LB
> UK

524 pH immersion probe, 502 pH immersion probe, 515 pH immersion probe, 521 pH immersion probe
  Dr W Ingold AG
  CH-8902 Urdorf
  Zurich
  Switzerland

Models 9160, 9170, 9150 and 9140, Industrial pH/redox instruments, intrinsically safe pH system, models 9188, 9189, System 18 modules, P96M pH controller, 2867, 7600, 7620 and 7670 pH electrode system, electrode cleaning system, models 7610/11, 7612/13
  Kent Industrial Measurements Ltd
  Oldends Lane
  Stonehouse
  Gloucestershire GL10 3TA
  UK

Electrical conductivity:
4080 panel-mounted EC meter with recorder
4080 HL panel-mounted EC meter with recorder and Hi–Lo alarm and relay output
4090 waterproof panel-mounted EC meter with recorder
4090 HL waterproof panel-mounted EC meter with recorder and Hi–Lo alarm and relay output

PHOX 50 conductivity recorder, panel-mounted
PHOX 55 weatherproof conductivity recorder panel-mounted
  PHOX Systems Ltd
  Ivel Road
  Shefford
  Bedfordshire SG17 5JU
  UK
  Jenway Ltd
  Gransmore Green
  Felsted
  Dunmow
  Essex CM6 3LB
  UK

WACA-120 conductivity meter with alarms and microprocessor
  Horiba International Corporation
  102 Duryea Avenue
  Irvine
  California 92714
  USA
  Sartec Ltd
  Bourne Industrial Estate
  Wrotham Road
  Borough Green
  Sevenoaks
  Kent TN15 8DG
  UK

Dissolved oxygen: 9090 water-resistant DO meter with recorder control, 9090 HL ditto with Hi–Lo alarm and relay output, DHOX60 dissolved oxygen recorders panel-mounted, DHOX65 weatherproof dissolved oxygen recorder, multi-channel dissolved oxygen control panel, WAXA-100 DO monitor with recorder
  PHOX Systems Ltd
  Ivel Road
  Shefford
  Bedfordshire SG17 5JU
  UK
  Jenway Ltd
  Gransmore Green
  Felsted
  Dunmow
  Essex CM6 3LB
  UK

WAXA-100 DO monitor with recorder
  Horiba International Corporation
  102 Duryea Avenue
  Irvine
  California 92714
  USA
  Sartec Ltd
  Bourne Industrial Estate
  Wrotham Road
  Borough Green
  Sevenoaks
  Kent TN15 8DG
  UK

Colour measurement: 8072 on-line colour monitor
  Kent Industrial Measurements Ltd
  Oldends Lane
  Stonehouse
  Gloucestershire GL10 3TA
  UK

Turbidity measurement: WATA-100 turbidity monitor
  Horiba International Corporation
  1021 Duryea Avenue
  Irvine
  California 92714
  USA
  Sartec Ltd
  Bourne Industrial Estate
  Wrotham Road
  Borough Green
  Sevenoaks
  Kent TN15 8DG
  UK

PHOX 70 turbidity recorder, panel-mounted, PHOX 75 weatherproof turbidity recorder, panel-mounted

Temperature measurement:
PHOX 30 temperature recorder, panel-mounted and PHOX 35 weatherproof temperature recorder, panel-mounted

Flow recorder:
PHOX 80 flow recorder, panel-mounted, series 155 flow recorder, panel on pedestal mounted

Sludge blanket detector PHOX 20, panel-mounted

Sludge level indicator: PHOX 25
  PHOX Systems Ltd
  Ivel Road
  Shefford
  Bedfordshire SG17 5JU
  UK

  Jenway Ltd
  Gransmore Green
  Felsted
  Dunmow
  Essex CM6 3LB
  UK

Free chlorine measurement:
Xertex PDS-924 chlorine measurement system

Ozone measurement:
Xertex PDS-940 ozone measurement system
  Delta Atlantic Corporation
  Marins Boulevard
  Hauppauge
  New York 11787
  USA

  Centronic Sales Ltd
  Centronic House
  King Henry's Drive
  New Addington
  Croydon
  Surrey CR9 0BG
  UK

Oz trameter
  Ozotech Ltd
  116 Station Road
  Burgess Hill
  West Sussex RH15 9EN
  UK

On-line chloride monitor models 8024 and 8034
On-line nitrate monitor model 8026, on-line phosphate monitors models 8063 and 8064,

on-line ion-selective monitors for fluoride, ammonia and nitrate model 8080, on-line silica monitor models 8065, 8061 and 8062 on-line sodium monitor model 8035, on-line total iron and copper monitors models 8066 and 8068, on-line total iron monitor for potable waters model 8076
  Kent Industrial Measurements Ltd
  Hanworth Lane
  Chertsey
  Surrey KT16 9LF
  UK

Cyanide analyser model HA 5400
  Iskra Commerce 61001
  Ljubljana
  PO Box 581
  Trg Revolucije 3
  Yugoslavia

  EDT Analytical Ltd
  14 Trading Estate Road
  London NW10 7LU
  UK

Digital ammonia monitor
  PHOX Systems Ltd
  Ivel Road
  Shefford
  Beds SG97 5JU
  UK

Section 10.2
On-line multiparameter instruments

Kent industrial monitoring system-19 comprising water-quality monitor 7975, with modules 9616 (pH), 9519 (EC), 7995 (susp. solids), 7996 (susp. solids), 9527 (dissolved oxygen), R1072 (temperature), monitor 7975. Kent 8080 series microprocessor, P105M circular chart recorder
  Kent Industrial Measurements Ltd
  Howard Road
  Eaton Socon
  St Neots
  Huntingdon
  Cambridgeshire PE19 3EU
  UK

  Kent Industrial Measurements Ltd
  Oldends Lane
  Stonehouse
  Gloucestershire GL10 3TA
  UK

Series 1800 continuously operating ion
selective electrode monitors
SA-9000 on-line process analyser
 Skalar BV
 Spinveld 62
 NL4815HT Breda
 PO Box 3237
 NL 4800 DE Breda
 Netherlands

 VW Scientific
 Outgang Lane
 Osbaldwick Industrial Estate
 York YO1 3UK
 UK

Series 8100 process analyser
 Dionex Corporation
 1228 Titan Way
 PO Box 3603
 Sunnyvale
 California 94088-3603
 USA

 Dionex (UK) Ltd
 Albany Park
 Camberley
 Surrey GU15 2PL
 UK

Trace metal analyses: PVA 2000 voltammetric
analyser
 Chemtronics Ltd
 Bentley
 Western Australia

 EDT Analytical Ltd
 14 Trading Estate Road
 London NW10 7LU
 UK

Organic pollution monitors: OA SA-1100
organic pollutant monitor
 Horiba International Corporation
 1021 Duryea Avenue
 Irvine
 California 92714
 USA

 Sartec Ltd
 Bourne Industrial Estate
 Wrotham Road
 Borough Green
 Sevenoaks
 Kent TN15 8DG
 UK

AODL-12 oil-in-water alarm, SOD-12 oil-in-
water alarm
 DKK Corporation
 13-14 4 Chrome
 Kita-Machi

 Kichijoji
 Masashino
 Tokyo
 Japan

 Sartec Ltd
 Bourne Industrial Estate
 Wrotham Road
 Borough Green
 Sevenoaks
 Kent TN15 8DG
 UK

Automated on-line UV pollution monitor
 Yorkshire Water Authority
 Kontron Instruments
 Blackmore Lane
 Croxley Centre
 Watford
 Hertfordshire WD1 8XQ
 UK

Telemetry equipment: Scadaflex system
 ATS (Telemetry) Ltd
 Burrell Road
 Haywards Heath
 Sussex
 UK

Chapter 11
On-site safety
instrumentation

Single-gas monitoring instruments
Crowcon 75 TC gas monitor, CO_2 in air, or H_2
in air/nitrogen, or H_2 in air/nitrogen or CH_4 in
air/nitrogen
 Crowcon
 Temple Road
 Cowley
 Oxford OX4 2EL
 UK

M-TEC combustible gas detector/alarms,
flammable gases, e.g. CH_4
 Datachannel (Gostex)
 2 Cochrane Close
 Crownhill
 Milton Keynes
 Buckinghamshire MK8 0AJ
 UK

Neotronics Digiflame 2000 flammable gas
detector, flammable gases, e.g. CH_4
 Neotronics Ltd
 Parsonage Road

Takeley
Bishops Stortford
Hertfordshire CH22 6PU
UK

Neotronics Neotox O_2 deficiency or
enrichment pocket-sized single-gas monitors,
O_2 deficiency or enrichment, CO, H_2S, SO_2 or
Cl_2
    Neotronics Ltd
    Parsonage Road
    Takeley
    Bishops Stortford
    Hertfordshire CH22 6PU
    UK

Multiple gas monitoring instruments:
Crowcon 84TR triple type gas tester:
simultaneous monitoring of (a) flammable
gases (calibrated for CH_4 and $C5-H_{12}$, (b)
H_2S, CO; (c) O_2 deficiency or enrichment
    Crowcon
    Temple Road
    Cowley
    Oxford OX4 2EL
    UK

Drager Triowarm three-in-one gas monitor,
simultaneous monitoring of (a) O_2, (b) H_2S or
CO, (c) explosive gases
    Drager Ltd
    The Willows
    Mark Road
    Hemel Hempstead
    Hertfordshire HP2 7BW
    UK

Neotronics Exolox range of gas monitors:
4-gas monitor model 50: (a) O_2 deficiency or
enrichment, (b) any two out of the three
following toxic gases: CO; H_2S; SO_2, (c)
flammable gases

Model 20/25/40 – 3-gas monitor: (a) O_2
deficiency or enrichment, (b) any one out of
the three following toxic gases: CO, H_2S, SO_2,
(c) flammable gases

Model 55 with GL15 logger – 4-gas monitor as
with model 50 but with real-time display of
results and full data logging facility

Models 60 and 70 Exotox atmosphere
monitor: up to 5-gas sensors (O_2, H_2S, SO_2,

CO and flammable gases) and temperature
and relative humidity monitoring

Exotox Model 75 ambilog ambient
temperature monitor: as above but allowing
up to 16 monitoring sessions totalling 30 hours
to be logged

IGD Sentinel 3-gas detector – 3-gas monitor:
O_2, H_2S, CH_4
    International Gas Detectors Ltd
    Sandbeck Way
    Wetherby
    West Yorkshire LS22 4DN
    UK

MSA Gasguard 742 – 3-gas monitor, O_2, CO,
CH_4
    MSA (Britain) Ltd
    East Shawhead
    Coatbridge
    Strathclyde ML5 4TD
    Scotland

GMI Landsurveyer 1 – land fill gas monitor:
3-gas monitor: O_2, CO_2, CH_4
    Gas Measurement Instruments Ltd
    Inchinnun Estate
    Renfrew
    Strathclyde PA4 9RG
    Scotland

Sieger Gas Leader monitor, 4-gas monitor for
CH_4, H_2S, O_2 deficiency, CO, SO_2 Cl_2, H_2,
NO and NO_2
    Sieger Ltd
    33 Nuffield Estate
    Poole
    Dorset BH17 7RZ
    UK

Standardization gas supplies:
Standard calibration mixtures available from
Neotronics and Crowcon. The latter supply
the following standards:
2.5% methane in air, 1.1% propane in air,
200 p.p.m. carbon monoxide in air, 25 p.p.m.
± 5 p.p.m. hydrogen sulphide in air,
150 p.p.m. carbon monoxide in air, 25 p.p.m.
± 5 p.p.m. hydrogen sulphide, 17% oxygen,
2.5% methane in nitrogen, 150 p.p.m. carbon
monoxide, 17% oxygen, 2.5% methane in
nitrogen

References and further reading

Adrian W.A. (1971). *Perkin Elmer Atomic Absorption Newsletter*, **10**, 96.

Aiginger H. and Wodbrauschek P. (1974). *Nucl. Inst. and Method*, **114**, 157.

Alekseva V. and Gd'dina T.S.A. (1950). *Zavod Lab.*, **16**, 35.

Allison J. Finnigan MAT IDT Publication No. 41 – the hows and whys of ion trapping.

Anderson M. (1988). *Liquid Chromatography Gas Chromatography*, **87**, 566.

Andrews J.N. *et al.* (1972). *Trans. Inst. Min. Metal*, B1 part 792–B 198 – B 209.

Babington R.S. US Patents 3,421, 692; 3,421, 699; 3,425,058, 3,425,059 and 3,504, 859.

Bajor M. and Bohling H. (1970). *Z. Anal. Chem.*, **249**, 190.

Bauer K. and Dreischer H. (1968). *Fortschr. Wasserchem. ihrer Grenzgeb*, **10**, 31.

Birnic S.E. (1988). *J. Automatic Chemistry*, **10**, 140.

Bishop P. Finnigan MAT IDT Publication No. 28, The ion-trap detector, universal and specific detection in one detector.

Bishop P. Finnigan MAT IDT Publication No. 36, The use of an ATD 50 GLC ion trap detector combination.

Bishop P. Finnigan MAT IDT Publication No. 42, low-cost mass spectrometer for GC.

Camp C. Finnigan MAT IDT Publication No. 23, Ion-trap advancements. Higher sensitivity and greater dynamic range with automatic gain control software.

Campbell C. and Evans S. Finnigan MAT IDT Publication No. 29. The ion-trap detector – the techniques and its application.

Campbell C. Finnigan MAT IDT Publication No. 15. The ion-trap detector for gas chromatography: technology and application.

Canate F., Rios A., Luque de Castro M.D. and Valcarcel M. (1987a). *Analyst* (London), **112**, 263

Canate F., Rios A., Luque de Castro M.D. and Valcarcel M. (1987b). *Analyst* (London), **112**, 267

Castle R.G. (1988) *J. Institute of Environmental Management*, 275, 2 June.

Cathrone B. and Fielding M. (1978). *Proc. Anal. Pro. Chem. Soc.* (London), **15**, 155.

Cognet L., Jacq P. and Mallevialle J. (1987) *Aqua*, No. 122.

Colby B.N. (1986). *Spectra.*, **10**, 49.

Commission for European Communities (1987). *Proposal for a Council Regulations (Euratom) Commission*, COM (87), 281 final.

Cothern C.R. *et al.* (1986). *Health Physics*, **50**, 33.

Criss J.W. and Birks L.S. (1968). *Anal. Chem.*, **40**, 1080.

Cross F.T. *et al.* (1984). *Health Physics*, **48**, 649.

Dadashev K.L.K. and Agamirova S.I. (1957) *Vos Issled Neft 2 nefte prod Razra botki profseeor pererabotki Neft obsled Zav Ustannov* No. 1, 81.

Dalton E.F. and Melanoski A.J. (1969). *J. Assoc. Official Analytical Chemists*, **52**, 1035.

Danyl F. and Nietsch B. (1952). *Mikrochemie Mikrochim. Acta.*, **39**, 333.

Das B.S. and Thomas G.H. (1978). *Analytical Chemistry*, **50**, 967.

Date A.R. and Gray A.L. (1981). *Analyst* (London), **106**, 1255.

Date A.R. and Gray A.L. (1983). *Spectrochim. Acta*, **38B**, 29.

Dawson R. and Pritchard R.G. (1978). *Marine Chemistry*, **6**, 17.

Dennis A.L. and Porter D.G. (1981). *J. Automatic Chemistry*, **2**, 134.

Douglas D.J., Quan E.S.K. and Smith R.G. (1983). *Spectrochim. Acta*, **38B**, 39.

Dunn B.P. and Stich H.F. (1976). *J. Fisheries Research Board* (Canada), **33**, 2040.

Eichelberg J.W. and Budd W.L. Finnigan MAT IDT Publication No. 47. Studies in mass spectrometry with the ion-trap detector.

Eicherberg J.W. and Slivon L.E. Finnigan MAT IDT Publication No. 48. Existence of self chemical ionisation in the ion-trap detector.

Ellis A.T. and Leyden D.E. Colorado State University private communication.

Engelhardt H. (1979). *High Performance Liquid Chromatography*. Springer, Berlin.

Engelhardt H. and Lillig B. (1985). *J. HRC & CC*, **8**, 531.

Engelhardt H. and Lillig B. (1986). *Chromatographia*, **21**, 136

Engelhardt H. and Neue U.D. (1982). *Chromatographia*, **15**, 403.

Engelhardt H., Klinkner R. and Lillig B. (1985). In *Kopplungsverfohren in der HPLC* (Engelhardt, H. and Hupe, K.P., eds.). GIT Verlag, Darmstadt.

Evans S. and Smith R.D. Finnigan MAT IDT Publication No. 11. The use of quadrupole and ion-trap mass spectrometers for the identification of air and water pollutants.

Evans S., Smith R.D. and Wellby S.K. (1984). *Intern. J. Mass Spectrometry and Ion Processes*, **60**, 239.

Fang Z., Ruzicka J. and Hansen E.H. (1984). *Anal. Chim. Acta.*, **164**, 23.

Federal Register (1979). **44**, 69466, 3 December.

Federal Register (1980). **45**, 33066, 19 May.

Federal Register (1984a). **49**, 38801, 1 October.

Federal Register (1984b). Method 624, **49**, 43234, 26 October.

Fernandez F.J., Lumas B. and Beaty M.M. (1980). *Atomic Spectroscopy*, **1** (2), March/April.

Finnigan MAT Application data sheet ADS11. Trace analysis of polychlorinated dibenzo-*p*-dioxins with the ion trap detectors.

Finnigan MAT Application data sheet ADS14. Gas chromatographic analysis of phthalate esters with the ion trap detector.

Finnigan MAT Application data sheet ADS22. Determination of polychlorinated biphenyls in industrial waste streams with ITD.

Finnigan MAT Application data sheet ADS7. Complex flavour analysis with the ion trap detector.

Finnigan MAT Application data sheet ADS8. Determination of Δ^9 carboxyl THC by GC/MS with the ion trap detector.

Finnigan MAT Application data sheet ADS9. Analysis of commonly abused drugs by GC/MS with the ion trap detector.

Finnigan MAT Application data sheet ADS10. Analysis of base neutrals by gC/MS with the ion trap detector.

Finnigan MAT Application data sheet ADS12. Analysis of coal extract with the ion trap detector.

Finnigan MAT Application data sheet ADS13. Gas chromatographic analysis of China White (Fentanyl) with the ion trap detector.

Finnigan MAT Application data sheet ADS14. High sensitivity full scan analysis of D3-11-NOR-9-Carboxyl-Δ^9-Tetrahydrocannabinol by the ITD.

Finnigan MAT Application data sheet ADS24. High sample levels with the ITD.

Finnigan MAT Application data sheet ADS26. Trace levels analysis of tetrachloro-dibenzo-*p*-dioxins with the ITD.

Finnigan MAT Application data sheet ADS27. Calibration of the ITD for quantification of tetrafluoro-1.4-dicyano benzene and 2.6 dichloro-4-nitroaniline.

Finnigan MAT Application data sheet ADS29. Dynamic range of the ITD.

Fisk J.F., Haeberer A.M. and Kovell S.P. (1986). *Spectra*, **10**, 22.

Freegorde M. (1971). *Laboratory Practice*, **20**, 35.

Frei R.W. and Lawrence J.F. (1981a). *Chemical Derivatization in Analytical Chemistry*, Vol. 1. Plenum Press, New York.

Frei R.W. and Lawrence J.F. (1981b). *Chemical Derivatization in Analytical Chemistry*, Vol. 2. Plenum Press, New York.

Friedman D. (1986). *Spectra.*, **10**, 40.

Furst P., Kruger C., Groekil H.A. and Groebel W. Finnigan MAT IDT 43. Determination of polychlorinated biphenyl substitute Ugilec (tetrachlorobenzyltoluenes) in fish.

Gardner W.S. and Lee G.F. (1973). *Environmental Science and Technology*, **7**, 719.

Genin E. Finnigan MAT IDT 53. Le Détecteur à Plegeage D'Ions de Chromatographie en Phase Gazeuse: Technologie et Applications

Godden R.G. and Thomerson D.R. (1980). *Analyst* (London), **105**, 1137.

Goulden P.D. and Brooksbank P. (1974). *Analytical Chemistry*, **46**, 1431.

Gray A.L. (1975). *Analyst* (London), **100**, 289.

Gray A.L. (1986). *Spectrochim. Acta.*, **41B**, 151.

Gray A.L. and Date A.R. (1983). *Analyst* (London), **108**, 1033.

Greenfield S., Jones I.L. and Berry C.T. (1964). *Analyst* (London), **89**, 713.

Grobenski Z., Lehmann R., Radzuik B. and Voellkoft U. (1984). Paper presented at the Pittsburgh Conference Atlantic City, March 5–9 1984. Application Study No. 656. The determination of trace metals in seawater using Zeeman graphite furnace AAS. Perkin-Elmer GmbH, Uberlingen, Germany.

Guerrieri F. and Bucci G. (1985). *Anal. Chim. Acta.*, **167**, 393.

Gunn A.M., Millard D.L. and Kirkbright G.F. (1978). *Analyst* (London), **103**, 1066.

Hallam C. and Thompson K.C. (1986). Determination of lead and cadmium in potable waters by atom trapping atomic absorption. Spectrometry Division Laboratory, Yorkshire Water Authority, Sheffield, UK.

Hargrave B.T. and Phillips G.A. (1975). *Environmental Pollution*, **8**, 193.

Haynes H. and Hodgkins J. Finnigan MAT IDT Publication No. 7. Packed column GC/MS with the ion trap detector.

Haynes W. (1978). *Perkin-Elmer Atomic Absorption Newsletter*, **17**, 49.

Hewlett Packard (1988). *Peak*, Autumn, 10.

Harlick G. *et al.* (1986). Winter Conference on Plasma Spectrochemistry, Hawaii, January. Paper 2.

Horzdorf C. and Janser G. (1984). *Anal. Chim. Acta.*, **165**, 201.

Houk R.S. (1980). *Anal. Chem.*, **52**, 2283.

ICRP (1985–86). *Measurement of alpha and beta activity of water and sewage sludge samples. The determination of Radon-222 and radium-226. The determination of uranium (including general X-ray fluorescent analysis)*. HMSO, London.

Jager J. and Fassovitzova B. (1968). *Chem. Listy.*, **62**, 216.

Karube I. and Suzuki M. (1986). *Biosensors*, **2**, 343.

Kelly P. Finnigan MAT IDT 21. Ion trap detector literature reference list.

Kelly P.E. Finnigan MAT IDT 10. New advances in the operation of the ion trap mass spectrometer.

Kennedy S. and Wall R. (1988). *Liquid Chromatography – Gas Chromatography*, **445**, 10.

Khesina A.Y. and Petrova J. (1973). *Spectroscopy USSR*, **18**, 622.

Kingston H.M. and Jassie L.B. (1986). *Anal. Chem.*, **58**, 2534.

Kirkbright G.F. and de Lima G.C. (1974). *Analyst* (London), **99**, 338.

Klesper E., Corwin A. and Turner D. (1962). *J. Organic Chemistry*, **27**, 700.

Knoth J. and Schwenke H. (1978). *Fresenius Z. Anal. Chem.*, **291**, 200.

Knoth J. and Schwenke H. (1980) *Fresenius Z. Anal. Chem.*, **201**, 7.

Krull I.S., ed. (1986). *Reaction Detection in Liquid Chromatography*. Marcel Dekker, New York.

Krull I.S., Bushell D.S., Schleicher R.G. and Smith S.B. (1986). *Analyst* (London), **111**, 345.

Kuchn D.G., Brandvig R.L., Lundeen D.C. and Jefferson R.H. (1986). International Laboratory, 82, September.

Kunte H. (1967). *Arch. Hyg. Bakt.*, **151**, 193.

Kurge K.H. (1959) *Morsk Flot.*, **19**, 31.

Laboratory News (1988). **8**, 11 July.

Later D., Bornhof D., Lee E., Henion J. and Wiedholt R. (1987). *Liquid Chromatography – Gas Chromatography*, 804.

Lawrence J.F. and Frei R.W. (1976). *Chemical Derivatization in L.C.*, Elsevier, Amsterdam.

Leger A., French Patent, I, 560, 544 (Paris Patent Office).

Leheir M., Finnigan MAT IDT 51. The use of the ITD a low cost GC/MS system for the identification of trace compounds.

Leonchenkova E.T. (1960). *Obogashch. Rud.*, **5**, 24.

Leoy E.M. (1971). *Water Research*, **5**, 723.

Lewis W.M. (1975). *Water Treatment and Examination*, **24**, 243.

Lichtenberg J.J., Bellar T.A. and Longbottom J.E. (1986). *Spectra*, **10**, 10.

Lloyd J.B.F. (1971a). *J. Forensic Science Society*, **11**, 83.

Lloyd J.B.F. (1971b). *J. Forensic Science Society*, **11**, 153.

Lloyd J.B.F. (1971c). *J. Forensic Science Society*, **11**, 235.

Markides K., Lee E., Bolick R. and Lee M. (1986). *Anal. Chem.*, **58**, 740.

Markos Varga G., Criky I. and Jonsson J.A. (1954). *Anal. Chem.*, **56**, 2066.

Marshall M.A. and Mottola H.A. (1985). *Anal. Chem.*, **57**, 729.

Matusiewicz H. and Barnes R.M. (1984). *Applied Spectroscopy*, **38**, 745.

Monarca S., Causey B.S. and Kirkbright G.F. (1974). *Water Research*, **13**, 503.

Muel B. and Lacrox G. (1960). *Bull. Chem. Soc.*, 2139.

Mullen W.H. and Vadgama P. (1986). *J. Appl. Bact.*, **61**, 181.

Nadkarni R.A. (1981). *American Laboratory*, **13**, 22, August.

Nadkarni R.A. (1984). *Anal. Chem.*, **56**, 2233.

Nelson A. (1985a). *Anal. Chim. Acta.*, **169**, 273.

Nelson A. (1985b). *Anal. Chim. Acta.*, **169**, 287.

Newstead S. (1988). *Radioactivity in drinking water*. Paper submitted to the Joint Working Group for radioactivity releases affecting the water industry, from HM Inspectorate of Pollution.

Nietsch B. (1954). *Angew Chem.*, **66**, 571.

Nietsch B. (1956a). *Gass. Wass. Warme*, **10**, 66.

Nietsch N. (1956b). *Mikrochim. Acta.*, 171.

Novotny M., Springston P.J. and Lee M. (1981). *Anal. Chem.*, **53**, 407A.

Ogan K., Katz E. and Slavin W. (1978). *J. Chromatographic Science*, **16**, 517.

Olsen E. Finnigan MAT IDT 35. Serially interfaced gas chromatography/ Fourier transform infrared spectrometer/ion trap mass spectrometer.

Olsen S.V., Pessenda L.R.C., Ruzicka J. and Hanson E.O. (1983). *Analyst* (London), **108**, 905.

Pahlavanpour B., Thompson M. and Thorne L. (1981). *Analyst* (London), **106**, 467.

Parker C.A. and Barnes W.J. (1960). *Analyst* (London), **85**, 3.

Parr Manual (1974). 207M Parr Instruments Co., 211 Fiftythird St., Moline, Illinois, 61265.

Pella P.A. and Dobbyn R.C. (1988). *Anal. Chem.*, **60**, 684.

Pickford C.J. and Brown R.M. (1986). *Spectrochim. Acta.*, **41B**, 183.

Place J.F., Sutherland R.M. and Dahne C. (1985). *Biosensors*, **1**, 321.

Pochkin Yu U. (1968). *Maisino. Oct. Kazan ed. Zh.*, **3**, 81.

Prange A., Knockel A. and Michaelis W. (1985). *Anal. Chim. Acta.*, **172**, 79.

Prange A., Knoth J., Stobel R.B., Bodekker H. and Kramer K. (1987). *Anal. Chim. Acta.*, **195**, 275.

Pritchard H.M. (1987). *J. American Water Works Association*, **79**, 159.

Quin R.Y., Min Y.Y., Xing L.Y., Qin L.R., Yu Z.C., Lan J.L., Jun P.H., Hui S.J., Chuan Z.Z. and Chau Z. (1986). *Spectra*, **10**, 56.

Ratnayake W.M.N. Finnigan MAT IDT 33. Mass spectra of fatty acid derivatives of isopropylidenes of novel glycerylethers of cod muscle and phenolic acetates obtained with the Finnigan MAT ion-trap detector.

Rence B.W. (1980). *J. Automatic Chemistry*, **105**, 1137.

Rence B.W. (1982). *J. Automatic Chemistry*, **4**, 61.

Reverz R. and Hasty E. (1987). Recovery study using an elevated pressure temperature microwave dissolution technique, presented at the 1987 Pittsberg Conference and Exposition on Analytical Chemistry and Applied Spectroscopy. March.

Richards J.M. and Bradford D.C. Finnigan MAT IDT 25. Development of a Curie Point Pyrolysis inlet for the Finnigan MAT Ion-Trap Detector.

Richards J.M., McClennan W.H., Burger J.A. and Menza H.H.C. Finnigan MAT IDT 56. Pyrolysis short-column GC/MS using the ITD and ITMS.

Righton M.J.G. and Watts C.D. (1986). Water Research Centre Report ER 1194-M. Identification of surfactants in water samples using sublation extraction and fast atom bombardment mass spectrometry. December.

Rivera J., Carxach S., Ventura F., Figueras A., Fraisse D and Des Salees G. (1985). FAB-CAD-MIKES Analysis of non-ionic surfactants in raw and drinking water. *Proceedings 10th International Mass Spectrometry Conference*, Swansea 9–13 September. (J.F.F. Todd, ed.). Wiley, Basingstoke.

Roederer J.E. and Bastiaans G.J. (1983). *Anal. Chem.*, **55**, 2333.

Rordorf B.F. Finnigan MAT IDT 13. An automated flow tube kinetics instrument with integrated GC-ITD analysis.

Rordorf B.F. Finnigan MAT IDT 14. Comparison of quantitative results for one analysis of 2, 3, 7, 8-TCDD in fish by 4500 Quadrupole MS and 705 ion trap detector.

Rowland A.P. (1986). *Analytical Proceedings* (London), **23**, 308, August.

Rubin R.B. and Heberling S.S. (1987). *International Laboratory*, **54**, September.

Rugin Y., Jian B., Hongjun P., Junhuis S. and Chungun Z. Chinese Academy of Environmental Sciences, Beijing, China. Finnigan MAT Application Report No. 215.

Ruz J., Torres A., Rios A., Luque de Castro M.D. and Valcarcel M. (1986). *J. Automatic Chemistry*, **8**, 70.

Ruzicka J. and Hansen E.A. (1975). *Anal. Chim. Acta.*, **78**, 145.

Ruzicka J. and Hansen E.H. (1978). *Anal. Chim. Acta.*, **99**, 37.

Rychkova V.I. (1969). *Elekt Sta Mosk.*, **40**, 76.

Salin E.D. and Harlick G. (1979). *Anal. Chem.*, **51**, 2284.

Salin E.D. and Szung R.L.A. (1984). *Anal. Chem.*, **56**, 2596.

Schalz L. and Attmann H.J. (1968). *Z. Analytische. Chem.*, **240**, 81.

Schlosser W. and Schwedt G. (1985). *Fresenius Z. Anal. Chem.*, **321**, 136.

Schnute W.C. Finnigan Corporation Applications Report No AR8018. Analysis of volatile organic compounds in industrial wastes by the Finnigan OWA GC/MS.

Schonmann M. and Kern H. (1981). *Varian Instrument Applications*, **15**, 6.

Schrader D.E., Volk L.M. and Covick L.A. (1983). *International Laboratory*, October.

Schwarz F.P. and Wasik S.P. (1976). *Anal. Chem.*, **48**, 524.

Schwenke H. and Knoth J. (1982). *Nucl. Meth.*, **193**, 239.

Scott R.H. (1974). *Anal. Chem.*, **46**, 75.

Shackleford W.W. and McGuire J.M. (1986). *Spectra*, **10**, 17.

Sharp B.L. (1984). The conespray nebulizer, British Technology Group, Patent assignment No. 8,432,338.

Shek W. Finnigan MAT IDT 4. Analysis of Polychlorinated biphenyls by gas chromatography – ion detector.

Sherman J. (1955). *Spectrochim. Acta*, **7**, 283.

Shklyar I.V. (1960). *Trudy Vses Nauchno-issled geologora Zved Int.* No. 155, 341.

Sim P., Elson C. and Quillaim M. (1988). *J. Chromatography*, **445**, 239.

Small H., Stevens T.S. and Bauman W.C. (1975). *Anal. Chem.*, **47**, 1801.

Sorrell R.K. and Reding R. (1979). *J. Chromatography*, **185**, 655.

Sorrell R.K., Dressman R.L. and McFarrer E.F. (1978). American Water Works Association water quality technology conference, Kansas City, Mo. December 5–7. American Water Works Association, Denver, Colorado, p. 3A–3.

Spackmann D.H., Stein W.H. and Moore S. (1958). *Anal. Chem.*, **30**, 1190.

Stafford G.C. Finnigan MAT IDT 16. Recent improvements in and analytical applications of advanced ion trap technology.

Stafford G.C. Finnigan MAT IDT 20. Advanced ion trap technology in an economical detector for GC.

Stafford G.C. Finnigan MAT IDT 24. The Finnigan MAT ion trap mass spectrometer (ITMS) – new development with ion-trap technology.

Stathan P.J. (1977). *Anal. Chem.*, **49**, 2149.

Steadman J. and Mantura R.F.C. Finnigan MAT IDT 8. The use of pollutant and biogenic markers as source discriminants of organic inputs to estuarine sediments.

Stepanova M.I., Il'ina R.H. and Shaposhnikov Y.K. (1972). *J. Anal. Chem. USSR.*, **27**, 1075.

Ster W. Finnigan MAT IDT 17. Analysis of contaminated water using the Finnigan MAT ion trap detector.

Stockwell P.P. (1979). In *Topics in Automatic Chemical Analysis*, Horwood, Chichester.

Stossel P.A. and Prange A. (1985). *Anal. Chem.*, **57**, 2880.

Suddendorf R.F. and Boyer K.W. (1978). *Anal. Chem.*, **50**, 1769.

Syka J.E.P. Finnigan MAT IDT 19. Positive ion chemical ionization with an ion trap mass spectrometer.

Tanaka K. and Fritz S.S. (1987). *Anal. Chem.*, **59**, 708.

Tanaka K. and Takeshita M. (1984). *Anal. Chim. Acta.*, **166**, 143.

Telliad W.A. (1986). *Spectra*, **10**, 4.

Thompson M. and Walsh J.N. (1983). In *A Handbook of Inductively Coupled Plasma Spectrometry*, Blackie, London and Glasgow, p. 55.

Thompson M., Pahlavanpour B. and Thorne L. (1981). *Analyst* (London), **106**, 467.

Thruston A.D. and Knight R. (1971). *Environmental Science and Technology*, **5**, 64

Tikkanen M.W. and Niemczyk T.M. (1984). *Anal. Chem.*, **56**, 1997.

Todd J., Mylchreest I., Berry T. and Games D. Finnigan MAT IDT 46. Supercritical chromatography mass spectrometry with an ion-trap detector.

Topping G. (1981). Marine Laboratory, Victoria Road, Aberdeen, Scotland.

Tretter H., Paui, G., Blum F. and Schreck H. (1985). *Fresenius Z. Anal. Chem.*, **321**, 650.

Tusck F., Lische P. and Chudova S. (1979). *Zeit für Wasser und Abwasser Forschung*, **12**, 242.

Uihlein M. and Schwab E. (1982). *Chromatographia*, **15**, 140.

Updike S.J. and Hicks G.P. (1977). *Nature* (London), **214**, 986.

US Environmental Protection Agency (1979). Industrial Environment Research Laboratory, Cincinnati, Ohio, p. 256 (32428).

US Environmental Protection Agency (1984). Method PA 600 14-54-017 March 1984. Environmental Monitoring and Support Laboratory, Cincinnati. The determination of manganese anions in water by ion chromatography method 300.0.

Vaughan C.G., Wheats B.B. and Whitehouse M.J.J. (1973). *J. Chromatography*, **28**, 203.

Wakeham S.G. (1977). *Environmental Science and Technology*, **11**, 272.

Wall, R.J. (1988). *Chromatography and Analysis*, J. Wiley & Sons.

Warburton G. and Millard B. (1984). *International Labmate*, **xii**, issue 7.

Watling R.J. (1977). *Anal. Chim. Acta.*, **94**, 181.

Welby J. Finnigan MAT IDT 1. Ion trap detector sensitivity using methyl stearate.

Welby J. Finnigan MAT IDT 6. Analysis of chlorinated pesticides using the ion trap detector.

Wendt R.H. and Fassel, V.A. (1965). *Anal. Chem.*, **37**, 920.

Westendorf R.G. (1986). Tekiman Company P.O. Box 371856, Cincinnati, OH 45223 – 11856, USA. Presented to the 1986 Water Quality Technology Conference of the American Water Works Association, Portland, Oregon, November.

Wilkes R. (1984). *Waste and Water Treatment*, 32, July.

World Health Organisation (1971). *International Standards for Drinking Water* (3rd edn). Geneva, p. 37.

Yaneda Y. and Horiuchi T. (1971). *Rev. Sci. Instr.*, **42**, 1069.

Yorkshire Water Authority, private communication.

Yost R.A., McClennan W. and Menzzelaar H.L.C. Finnigan MAT IDT 22. Enhanced full scan sensitivity and dynamic range in Finnigan MAT ion trap detector with new automatic gain control software.

Yudilevich M.M. (1954). *Luminescence methods for the determination of content of mineral oils in water*, Moscow, Gosudarst, Energet, Izdatel.

Yudilevich M.M. (1960). Luminescent method and apparatus for the analysis of water–oil emulsions Methody hyuminsteestsentu Analiza, *Minsk Akad. Nauk. Belorum*, SSR, Sbhornik 87.

Yudilevich M.M. (1964). *Maj. Peredoye Metody Teknol. 2 Kontrolya Proizv.*, 280.

Yudilevich, M.M. (1966). *Vodopodgotovka Vad. Rezhim. Khim. Kontrol Parosilovykh. Ustanvokakh*, S.B., Statei No. 2, 173.

Index